TIME AND TIMELESSNESS

TIME AND TIMELESSNESS

or

The Varieties of Temporal Experience

(A Psychoanalytic Inquiry)

PETER HARTOCOLLIS, M.D., Ph.D.

INTERNATIONAL UNIVERSITIES PRESS, INC.
New York

Library of Congress Cataloging in Publication Data

Hartocollis, Peter.
Time and timelessness, or, The varieties of temporal experience.

Bibliography: p.
Includes index.
1. Time—Psychological aspects. 2. Psychoanalysis.
I. Title. II. Title: Varieties of temporal experience.
[DNLM: 1. Time. 2. Time perception. 3. Psychoanalytic theory. BF 468 H334t]
BF468.H37 1983 153.7'53 83-237
ISBN 0-8236-6545-3

Manufactured in the United States of America.

No peevish winter wind shall chill
No sullen tropic sun shall wither
The roses in the rose-garden which is ours and ours only

From *A Dedication to My Wife* by T. S. Elliot

CONTENTS

PROLOGUE

PLATO TRIED TO DENY IT in his *Timaeus*. St. Augustine obsessed about it in his *Confessions*. Thomas Mann described it as both real—the indispensable vehicle of narration and music—and as a metaphor, a vast, misty ocean around his *Magic Mountain's* lonely, tuberculous souls. Ancient Zen Buddhists, and more recently Bergson and Heidegger, identified it with the self or being.

It has always been an interesting, mysterious subject, never taken lightly or forgotten. Time figures ever more prominently in our lives and in current literature, scientific as well as popular. People seem to be paying more attention to time these days. It is a contemporary phenomenon that we all are more aware of the changes identified with the concept of time, cultural and socio-economic changes brought about by breathtaking technological advances, fast, readily accessible means of transportation, and a wide, more or less reliable but far from reassuring dissemination of information and ideas. Change or motion is the main ingredient of time. Unless one is mentally retarded or in solitary confinement, one cannot fail to notice it.

Another reason for our greater awareness of time these days is that we are more anxious about things, more unsure about our personal identity, more concerned about our physical and social image, more afraid of aging and dying. An excess of narcissism and borderline personality features in all of us may explain why we are so concerned about time—not only more aware of its passage, but also more bent to do something about it, to slow it down or speed it up, to take advantage of it, or to merely deny it.

Time is influenced by one's normal or pathological state of being and character structure, by states of emergency, drugs, meditation, altered states of consciousness. In the following pages, I shall try to analyze all these states or conditions in terms of the experience of time, describing along the way its obligatory coordinates of self and affect as they define the quality of the experience and as they themselves are qualified by it.

This is a book about time and the self—the sense of time as experienced by the self both in normality and pathology. I might add, this is a book about time as a psychoanalyst views it. It is based on my psychoanalytic experience and readings. But I have borrowed freely from the literature of other schools and disciplines, primarily from phenomenology, beginning with the writings of Bergson, for whom also the concept of the self is central to the understanding of time, the experience of time as duration. Finally, let me confess that a great deal of what I am going to say about the subject is based on self-analysis—to put it bluntly, on introspection.

I wish to acknowledge my debt to all those who helped me personally with this book, beginning with my psychoanalytic colleagues in Topeka and elsewhere in the world, which reminds me of what Winnicott (1965) wrote on a similar occasion: "I have grown up as a member of this group, and after so many years of inter-relating it is now impossible for me to know what I have learned and what I have contributed. The writings of any one of us must be to some extent plagiaristic" (p. 11).

I thank The Menninger Foundation for its generous, timeless and selfless support, personified in Cotter Hirschberg, Chief of Professional Services, trusted and trustful friend, fellow analyst, scholar and humanist. I am grateful to my secretary, Carole Dillingham, whose accurate work has been an important part of each of the book's chapters at its original coming to life.

I also wish to acknowledge my debt to all those authors and publishers, who have allowed me to quote from their papers, journals, and books, in particular: International Universities Press, the *Bulletin of the Menninger Clinic,* the *Journal of the American Psychoanalytic Association,* the *International Journal of Psycho-Analysis,* and the *Psychoanalytic Quarterly.*

I

TIME AND CONSCIOUSNESS

TIME IS CONCEIVABLE in two broad ways: as subjective experience, what is known as psychological or absolute time, and as objective, clock, calendar, or mathematical time. Time in its experiential sense is the unifying element of consciousness, a process that ascribes unity to the perception of the self in a world of constant change. Time provides a sense of identity to the experience of discontinuous emotional states and to the perception of discontinuity in change. Time is the yardstick of the experience of change. The unifying property of time makes what was before, what is now, and what will be next appear related in a causal, personal way.

Time and Change: The Specious Present

As Janet (1928) recalled in one of his lectures on time, whenever Bergson spoke on the subject he would finish by saying: "there is no other thing in time than change itself" (p. 84). Likewise, William James (1890) wrote: "Awareness of change is . . . the condition on which our perception of time's flow depends" (p. 620). Such awareness defines what has been possible within the so-called specious present.[1] As James put it:

> Objects fade out of consciousness slowly. If the present thought is of ABCDEFG, the next one will be of BCDEFGH, and the one after that of CDEFGHI—the lingerings of the past

[1]The term was coined by the English psychologist E. R. Clay (1882), who wrote: "The present to which the datum refers is really a part of the past—a recent past—delusively given as being a time that intervenes between the past and the future. Let it be named the specious present, and let the past, that is given by being the past, be known as the obvious past" (p. 167).

> dropping successively away, and the incomings of the future making up the loss. These lingerings of old objects, these incomings of new, are the germs of memory and expectation, the retrospective and the prospective sense of time [p. 606].

Hughlings Jackson (1874–1876), the pioneer British neurologist whose ideas about epilepsy and the workings of the brain influenced Freud in more than one way, had this to say about time and the specious present:

> Time is required for consciousness. On that matter I will quote Spencer. . . 'A subjective state becomes recognizable as such, only when it has an appreciable duration: it must fill some space in a series of states, otherwise it is not known as present' [p. 205].

Eissler (1955) tells us that phenomenologist Edmund Husserl described time as "a stream of experiences with an infinite chain of *now*-points (impressions), each of them surrounded, so to speak, by a retention (a now-point just passed) and a protension (an expectation of a now-point which is still in the future but which becomes a now-point in the present" (p. 272).

As Thomas Mann (1924) put it:

> [Time] is the medium of narration, as it is the medium of life. Both are inextricably bound up with it, as inextricably as are bodies in space. Similarly, time is the medium of music; music divides, measures, articulates time, and can shorten it, yet enhance its value, both at once. Thus music and narration are alike, in that they can only present themselves as a flowing, as a succession in time, as one thing after another; and both differ from the plastic arts, which are complete in the present, and unrelated to time save as all bodies are, whereas narration—like music—even if it should try to be completely present at any given moment, would need time to do it in [p. 541].

The phenomenon of the specious present accounts for the sense of duration, what James (1890) also called "primary memory"—the overlap between fading impressions and fresh ones. To put it differently, the specious present is the length of time during which consecutive events are present together in consciousness. James pointed out that the length of the specious present and the number of discernible gradations within it, including the degree of vividness of the experience, can be altered by such factors as fatigue and drugs. Indeed, when tired or sleepy, even though aware of what one reads or hears, one loses this awareness so fast that attempting to respond would be likely to result in a parade of logical, perhaps, but meaningless words — words or phrases out of context, more or less like the speech of a person afflicted by Korsakoff's psychosis. In contrast, users of psychedelic drugs report a widening of consciousness, an intensification of the sense of present that creates the illusion of eternity or timelessness. Mystics and people in love describe similar experiences.

Time is the experience of change or movement, but only if something is perceived or experienced as unchanged. As the perception of change is predicated on the perception of a stationary background, in order for the perception of time to occur it is necessary to experience a stable or static self: time is the experience of change with reference to a personal self. To the extent that attention is on the self, the perception of change around it is accentuated. Time appears to move fast, urgently, relentlessly, or to slow down and stop — time weighs heavy on one's self as an external thing; time becomes of essence. To the extent that attention is away from the self, change is perceived but little; the time sense merges into the essence of one's own self — time ceases to exist.

The experience or sense of time, and later the perception of time as an attribute of objective reality, is a function of consciousness.[2] It grows along with consciousness, beginning with the differentiation of the self from the object world and the awareness that objects about oneself move or change. The baby's awareness of the flow of time and his own position in time's continuum is not, however, based on the cognition of external reality and movement as such, but on the perception of

[2] As Wittgenstein (1921) put it, "Space, time, and colour (being coloured) are forms of objects" (p. 11).

subjective reality in terms of tension-arousing, libidinal, or aggressive drives. What gradually establishes the sense of time as duration, and more or less coincidentally as temporal perspective, is the felt inadequacy of the self in terms of growing unpleasure and the awareness of the possibility that the need-fulfilling object—mother—may or may not come.

Time and the Unconscious

The idea that time is a modality of consciousness has been an integral part, if not an outgrowth, of the psychoanalytic theory of the unconscious as a dynamic psychological system that operates outside of the rules of Aristotelian logic. According to Freud (1915), "The processes of the system *Ucs.* [unconscious] are *timeless*; i.e. they are not ordered temporally, are not altered by the passage of time; they have no reference to time at all. Reference to time is bound up, once again, with the work of the system *Cs.*" (p. 187). Returning to the subject many years later, and with obvious reference to Kant's view of time, Freud (1933) elaborated:

> . . . we perceive with surprise an exception to the philosophical theorem that space and time are necessary forms of our mental acts. There is nothing in the id [the system *Ucs.*] that corresponds to the idea of time; there is no recognition of the passage of time, and—a thing that is most remarkable and awaits consideration in philosophical thought—no alteration in its mental processes is produced by the passage of time. Wishful impulses which have never passed beyond the id, but impressions, too, which have been sunk into the id by repression, are virtually immortal; after the passage of decades they behave as though they had just occurred. They can only be recognized as belonging to the past, can only lose their importance and be deprived of their cathexis of energy, when they have been made conscious by the work of analysis. . . [p. 74].

Freud (1920) attributed the origin of the notion of time to "the method of working of the system *Pcpt.-Cs.* [perceptual consciousness] and to... a [corresponding] perception on its own part of that method of working" (p. 28). As Freud (1925b) himself clarified in discussing the process of memory:

> It is as though the unconscious stretches out feelers, through the medium of the system *Pcpt.-Cs.*, towards the external world and hastily withdraws them as soon as they have sampled the excitations coming from it... This discontinuous method of functioning of the system *Pcpt.-Cs.* lies at the bottom of the origin of the concept of time [p. 231].

Freud's ideas concerning the psychological origin of time are remarkably similar to those of William James (as cited by Eisendrath, 1971), who suggested that perceptions or experience come "in quanta" and that the concept of time develops out of "'incessant impulses of voluntary reinforcement' in maintaining attention with effort" (p. 143). They are also reminiscent of Bergson's theory of the self as a dynamic entity made up of a qualitative multiplicity of impressions that are experienced as discontinuous duration, whose elements permeate one another and, as they are projected outward, create the impression of time as a sequence in space, as a continuum of events that lie side by side. Freud's, James's and Bergson's ideas on the subject seem to derive from a common background of nineteenth-century psychological theories, which stressed the role of active attention and will in the appreciation of time as duration, and from newly advanced concepts in physics—relativity and quantum theories. Such ideas reflect also longstanding psychophysiological speculations, according to which the experience of time is based on some sort of internal biological process, what Ornstein (1969) has described as a "pulse-dispensing mechanism which delivers internal time signals... [and] is identified either with a specific periodicity... called a time quantum or with a specific bodily process called the biological clock" (p. 25).

Neurophysiologists, on the other hand, describe consciousness—on which self-consciousness and the sense of time as duration or perspective, of course, depend—in terms of vigilance, alertness, wakefulness, and awareness (Eccles, 1966). Traditionally, consciousness has been attributed to the cerebral cortex as its characteristic overall function or property—i.e., as a property of the mind. More recently, however, this notion has been abandoned or, rather, modified by the discovery of the significance of the reticular formation or gray matter of the brain stem. It is now believed that the ascending reticular activating system provides the basis for the phenomenon of consciousness in terms of the so-called alerting response, known also as activation reaction, desynchronization response, or blocking reaction. The reticular formation and the reticular activating system are predominantly concerned with the state of wakefulness, while the cerebral cortex is more closely connected with the content of consciousness. It can thus be said that the cortical functions are superimposed upon the field of consciousness as to an infrastructure, the cortical superstructure providing a conceptual model of the world, a model that is probably phylogenetically determined.

Time as Duration

Time as a conscious, subjective phenomenon exists in two ways: as a sense of duration, and as tense or perspective. Time as duration is a sense or feeling. The sense or feeling of duration is based on the ability to experience one's self and the world of internal objects as continuous, as flowing together, synchronously, simultaneously, harmoniously; or apart, disjointly, incongruously, out of step with each other, fast or slowly, or as slowly as to be hardly moving.

The sense of time as duration is more elementary, developing before that of time as future, past, or present. More specifically, time as perspective is a derivative of the sense of duration and as such more conceptual than intuitive or feelinglike in nature. In Kant's terminology, time as duration is a form of intuition; time as perspective is a conceptual category, a modality of experience.

Elementary and precocious as it may be, however, the sense of duration is far from simple. The awareness of time as duration presupposes the development of more elementary notions or "intuitions," as Piaget in his Kantian orientation calls them—those of space, simultaneity, speed (or movement), and succession. Taking as his point of departure his definition of space as "a coordination of changes of position," Piaget (1970) describes time "as a coordination of movements or speeds" (p. 59). But, even though dealing with it psychologically, such a definition fails to convey the sense of time as an experience, ignoring the fact that time is not only an abstraction but also a feeling. And that applies to time as duration as well as perspective.

Time in the sense of duration can be defined as the experience of change within the limits of one's attention span. It is a subjective psychological phenomenon that varies from individual to individual and in the same individual for different mental states as they are determined by internal and external factors. Such variables are a matter of speculation and experimentation. It should also be noted that the phenomenon is subject to different rules in the case of very brief periods of time as measured by the clock, when the objective length of time covers only a few seconds or minutes (which is the length of time most laboratory studies on the estimation of time deal with) rather than the temporal units of ordinary, normal or abnormal experience, i.e., hours, days, years. Some of the factors accounting for the variability in the estimation of the experience of duration are the amount of effort exercised during a given time-span, the number of discrete acts or events contained in that period, and the pleasant or unpleasant nature of the content of that period (Doob, 1971). All these variables obviously depend on the individual's psychophysical condition.

The sense of duration is attributed to the external world at large and perceived as an extrasensory, pervasive, interrelational quality of objects, describable in terms of its apparent speed or length of trajectory. It is conceptualized as time. Indeed, the adult sense of duration involves something more basic than the experience of time as flowing, as unfolding or passing, fast or slowly. That is the feeling that objects—persons, things, institutions, ideas—have a lasting quality, exist in a temporal context. It is the feeling that things endure, that something has existed and will continue to exist beyond the experiential moment of recognition, the feeling that a person is of a certain age and therefore shares a comparable childhood history and a destiny marked by death, the feeling that a town or a church or a landscape was there and will be there whether we are there or not. Such is the essence of psychological time, the experience of time as a dimension of existence, the awareness of time as a quality of the object world. This feeling of things as enduring, of things as lasting longer than our span of attention, is an indispensable component of consciousness. Without it, consciousness is conceivable merely as a matter of animal sensitivity rather than human sensibility or meaning—which is another way of saying that time is a fundamental dimension of consciousness.[3]

The ontogenetic origins of the feeling or, in Kant's terms, intuition of time as duration, can be traced into the early developmental period that results in the establishment of object constancy, the ability to identify oneself as separate from objects and objects as different from each other in their own terms rather than as extensions of one's instinctual drives, as objects defined by one's needs and feelings. As Kubie (1954) put it: "Consciousness of anything implies a capacity to differentiate between perceptual processes which pertain to the self (the 'I'), and those pertaining to the rest of the world (the 'non-I')" (p. 465).

Time as Perspective

> Like a recurrent dream, I recall experiencing over and again the same awesome feeling of discovery as a child in my latency: drinking milk from a seemingly plentiful bowl and suddenly left staring at the empty round bottom of the bowl, holding it tightly in my raised hands, my lips still wet—a screen memory for the frustration of a now-plentiful-now-empty breast? I recall a similar experience as I watched the sun go down over the clear hori-

[3] Boring (1933) considered time or propensity (Titchener's term for duration) as one of the four major dimensions of consciousness (the other three being quality, intensity, and extensity). Holding that such dimensions are heuristic concepts rather than entities to be discovered, he saw them as continua on which a complete account of mental organization may be displayed.

> zon, a fleeing, ephemeral, transitional state of now-is-now-is-not; and, again, high up on a tree, ready to jump to the ground but holding back, forcing time to stand still by holding onto my place high up on a limb, when I should have been flying into the open space below, a second, two seconds, three seconds ago. Obsessive as they may sound clinically, I take these to be moments of a child's early search and insight into the nature of self-awareness, connecting the passage of time with the existence of the self as a conscious, autonomous, lonely but willful entity.

The feeling of duration and, in general, the consciousness of the world as time-bound, creates the experiences of before and after, of beginning and end, which result in the notion that all things and experiences as such have a past and a future, a beginning and an end point. As they refer to conceptual issues encompassed into the overall concept of life or existence, the notions of beginning and end become converted into what in essence are institutional notions—those of birth, growth, progress, generation, conservation, decay, death. These notions differ in their emotional as well as semantic definition according to the particular sociocultural context in which they are held (Gorman and Wessman, 1977).

The intuition of beginning and end becomes a personal, acute, emotional matter when its focus shifts from the object to the self, from the experienced object to the experiencing subject, to the individual as an experiencing object—a situation that defines self-awareness. The feeling of duration is dependent on the capacities to remember and to wait. These capacities, in association with the fulfillment or frustration of object-oriented needs and wishes, create the experience of time as tense or perspective, past and future.

> I do not remember how old I was, no older than five or six, standing at a busy corner of the town in the late afternoon, next to a large package or a small piece of furniture, when a couple of women stood in front of me and asked what was I doing there alone, did I need any help. I remember saying to them, "No, thank you, I am just waiting for my mommy;" and I remember them saying to each other, "Isn't he nice; let's wait with him until

> his mommy comes." Which they did, I cannot tell for how long, for I do not remember my mother coming. I just remember waiting for her, worrying about the two kind ladies who were standing there waiting with me.

Without the concept of time one's experience would remain discontinuous; growing objects around oneself would acquire new identities, become new persons and new products; mother would no longer be recognized as mother after one could fetch food for oneself; a ripe red apple would be a different fruit, not a mature apple when compared to the small green sour fruit that was.

The development of the experience of time as perspective is intimately related to the development of affects and conscience—the superego. By warning the ego of the dangerous consequences of instinctual wishes, the superego assists in the development of the ability to anticipate the future; and by condemning the ego's failure to resist such wishes, it assists in the development of the sense of an irrevocable past. The sense of time as either duration or tense varies according to the affective state of the individual, reflecting the relative state of adequacy of his ego or the integrity of his self.

Even though the experience of time is ubiquitous and in its basic aspects—time as duration and time as perspective—nearly universal, its study would be incomplete without considering how people succeed or fail to integrate it in their lives and circumstances. Age, intelligence, culture, social class are responsible for wide variations in the experience as well as the conceptualization of time. The greatest variations seem to be in the conceptualization of time as future and in the experiential equivalents of future time, anticipation, and anxiety in the face of the uncertain, the ambiguous, and the unknown—all of which defines, of course, the present and, experientially, is but the present. For one cannot experience the future except in one's fantasy or in one's wish and ability to cope with the present. The same may be said about the past. But, here again, there is no way to conceptualize the past without the future, either as a lively anticipation that saves the past from extinction and in effect mistakes it for the present, or as a dreadful eventuality that throws one back to the past as to a fatal, irresistible force, whose magnetic field dominates the present.

Time as perspective is an image, a concept, or both. What is present and what is future and what is past are but elaborations of the existential question of what is happening or not happening to the self in relation to the world of love objects—objects which we cathect, reflecting our internalized objects, people and what they have come to represent in terms of personal memories, values, and aspirations. The present is our wishes and their shape; and what the future holds that is meaningful stands for the present shapelessness of our wishes—the wish to reshape or keep shaping our wishes. The future is grounded in the present and the present is grounded in the past, the world of internal objects, the internalized object world that defines our being. The unifying concept of time—time as perspective—makes what was before, what is now, and what will be next appear related in a causal, personal, responsible way, allowing for planning and social adaptation.

II
TIME AND THE SELF: FROM BERGSON AND HEIDEGGER TO PIAGET

SPEAKING ABOUT TIME as experience, one can hardly avoid using the language of phenomenology and existentialism, "being" and "becoming" and "essence," for the study of experience has been pursued most explicitly by phenomenologists and existentialists—Hegel, Kierkegaard, Husserl, Heidegger, and Binswager, not to speak of the ancient Greeks, pre-Socratic philosophers such as Heraclitus, Embedocles, and Parmenides, who tried to define the nature of the universe on the basis of their own inner experience and intuition. Bergson may not be formally counted among them, but his evolutionist philosophy and his belief that intuition was more effective than logic in apprehending the true nature of existence had a very broad appeal, influencing such major figures in world literature as George Bernard Shaw, André Malraux, Nikos Kazantzakis, and a score of prominent philosophers and psychologists such as Alfred North Whitehead, Karl Jaspers, and Jean Piaget.

Time as Experience

As is true of most great philosophers across the centuries, Bergson was as much a psychologist as he was a logician and metaphysicist. In fact, his first major work, *Time and Free Will,* published in 1889 under the title *Essais Sur les Données Immédiates de la Conscience,* is an introspectionist treatise on the subject of time as experience or duration, and the self or ego as the seat of consciousness. He considered time and consciousness indispensable for the understanding of human nature and reality in general, psychological and physical as well.

Bergson (1889) pointed out that time is of two kinds: "abstract," as conceived by mathematics, physics, science in general; and "real concrete" time, apprehended as a "feeling of duration." Time as duration derives from the awareness of internal psychic states, ego states or experiences of the self, which compose distinct units of variable intensity but, when considered in their multiplicity, "they unfold in time and constitute *duration*" (p. 224). The feeling of duration makes these ego states appear to have a certain unity, to be internally consistent as if determining one another. The outcome is the sense of self as a unitary experience that is endowed with free will.

Time as an inner experience is, according to Bergson (1889), "a qualitative multiplicity, with no likeness to number; an organic evolution which is yet not an increasing quantity; a pure heterogeneity within which there are no distinct qualities" (p. 226). When this internal experience is projected outward, it defines the experience of the present, which is no more than the experience of simultaneity of objects or external events. At a given moment, the self experiences a multiplicity of simultaneous positions, which are independent of the simultaneities that have preceded them. In other words, it is only the self that endures and not the external objects or positions. External things move or change, but they do not succeed each other as a unitary sequence of events except insofar as we project onto them our own inner sense of time, the consciousness of our own self as enduring. In Bergson's words:

> To the simultaneities, which constitute the external world, and, although distinct, succeed one another *for our consciousness,* we attribute succession in themselves. Hence the idea that things *endure* as we do ourselves and that time may be brought within space [pp. 227–228].

Objective experience—the way the external world looks—is characterized by extensity and duration. The sense or quality of extensity defines the experience of space or, rather, the notion of space, while duration defines the sense or notion of time. It is the notion of space that makes the world appear homogeneous, divisible in equal periods, and hence measurable; while the notion of time contributes the sense of seriation, succession, or progress in change. Time without space or extensity is the experience of the self "as a wholly qualitative multiplicity, an absolute heterogeneity of elements which pass over into one another" (Bergson, 1889, p. 229). It is the experience of pure self, the self without objects, that is identified as eternity and may be pleasurable or unpleasurable, depending on the intensity of the sensory stimulation that impinges on the awareness of the self.

Inner time or duration is virtually indistinguishable from the awareness of the self, the experience of the self as an enduring, unitary entity that is constantly becoming. As duration is generated by moments that contain one another, the self is made up of inner states that are determined by each other, that succeed one another in a generative process rather than juxtaposing each other in a homogeneous space, inner states that envelop each other like onion rings or Russian dolls, a self that is multiple and heterogeneous, experienced as intensity or quality rather than measurable quantity. The self, like real inner time, is heterogeneous or discontinuous; but as it unfolds in time and not in space, it is experienced as a qualitative multiplicity that endures as a process, "a pure heterogeneity within which there are not distinct qualities" (Bergson, 1889, p. 226).

The inner states of the self, accounting for the experience of time as duration, permeate one another; they lie inside each other rather than side by side. Space, on the other hand, is the intuition of a homogeneous medium which enables the self to experience objects and external events in a predictable, quantifiable way that allows for control and the concept of causation. When the inner sense of time as duration is externalized, that is, when it is projected in space and experienced as part of the object world, it becomes time as tense or perspective; it is converted into the continuum of time that is defined as present, past, and future. In short, time as tense or perspective is outer time, time experienced as space or conceptualized in spatial terms. It defines the world of object relations, the relation of objects to other objects, or the self as a social entity.

For Bergson time as duration is primary, an intuition or inner sense that precedes that of space, rather than the two being equivalent structures of the mind and presumably originating together in experience, as Kant would have it. The notion of space is based on the experience of negation and sequence—negation in the sense that no two objects can occupy the same position at the same time, and sequence in the sense that things are perceived as lying side by side. Experiences of the external world, accounting for the notion of space, are secondary to the experience of one's inner states or experiences of the self, accounting for the notion of time. Time is conceptualized as a homogeneous medium in which distinct states of consciousness are arranged side by side as in space, only because one tends to project one's experience of duration into space and thus perceive time in spatial terms, as a discrete multiplicity. Self-images are projected along a continuum of space that is arranged according to three discrete areas of experience: possible or potential (conceptualized as future), becoming or taking place (conceptualized as present), and actual or having been (conceptualized as past).

Bergson saw time and space as inherent structures of the mind which emerge into consciousness under the impact of experience as generated by appropriate inner and outer stimuli. He was a convinced structuralist, as were most psychologists around the turn of the century. As Bergson (1889) himself put it: "Modern psychology seems to us particularly concerned to prove that we perceive things through the medium of certain forms, borrowed from our own constitution" (p. 222).

Understandably, he attributed this trend to Kant, and contrasted it to that of the empiricists who, led by William James, were just beginning to gain ground in psychology. Bergson (1889) went on to say that "the most obvious states of the ego itself, which we believe that we grasp directly, are . . . perceived through the medium of certain forms borrowed from the external world, which thus gives back what we have lent it" (p. 223). What characterizes space is extensivity or "extensive homogeneity," and what characterizes time is intensity or qualitative heterogeneity. Objects, being separate from one another and from the self, derive their separateness or, in Bergson's (1889) term, exteriority, "from the homogeneity of a medium which inserts intervals between them and sets off their outlines: but states of consciousness, even when successive, permeate one another, and in the simplest of them the whole soul can be reflected" (p. 98). There is an interpenetration, an interdependence of representations. Each one of them stands for the whole and cannot be distinguished or stand out alone, except by means of abstract thought. When, however, we think of time, we fall back into the spatial experience of succession, perceiving our states of consciousness no longer in one another, but alongside one another; we think of time as of space, conceptualizing duration in terms of extensity so that succession takes the form of a continuum, a line or chain of events that just touch each other rather than being within each other. Such a mental image creates the impression of a before and an after, the notions of earlier and later, which are no longer successive but simultaneous, mutually exclusive. Hence the dilemma between the notion of the irreversibility of time, on the one hand, and of time as a continuum, as a homogeneous multiplicity that allows for reversibility, on the other, as with the experience of space. So long as it is not conceived in spatial terms or as altered states of consciousness, however, time as the pure experience of duration that consists in permeable or interpenetrable heterogeneous states of consciousness is, in effect, reversible—time as experienced in regression or as a widening of consciousness. Bergson (1889) put all this graphically in analyzing one's perception of the movement of a pendulum and the resulting experience of time as duration:

> When I follow my eyes on the dial of a clock, the movement of the hand which corresponds to the oscillations of the pendulum, I do not measure duration, as seems to be thought; I merely count simultaneities, which is very different. Outside of me, in space, there is never more than a single position of the hand and the pendulum, for nothing is left of the past positions. Within myself a process of organization or interpenetration of conscious states is going on, which constitutes true duration. It is because I *endure* in this way that I picture to myself what I call the past oscillations of the pendulum at the same time as I perceive the present oscillation. . . . Owing to the fact that our consciousness has organized

> them [oscillations] as a whole in memory, they are first preserved and afterwards disposed in a series: in a word, we create for them a fourth dimension of space, which we call homogeneous time, and which enables the movement of the pendulum, although taking place at one spot, to be continually set in juxtaposition to itself. . . . There is a real space, without duration, in which phenomena appear and disappear simultaneously with our states of consciousness. There is a real duration, the heterogeneous moments of which permeate one another; each moment, however, can be brought into relation with a state of the external world which is contemporaneous with it, and can be separated from the other moments in consequence of this very process. The comparison of these two realities gives rise to a symbolical representation of duration, derived from space. Duration thus assumes the illusory form of a homogeneous medium, and the connecting link between these two terms, space and duration, is simultaneity, which might be defined as the intersection of time and space [pp. 107–110].

What helps establish time as a projection into space is the concept of motion, the fact that movement takes place in space, the successive positions of a moving body in space. It is movement or velocity that helps duration assume the form of a homogeneous medium, that makes us perceive duration in spatial, linear terms.

Time as Evolutionary Process

Time in its experience as duration is a dominant theme in Bergson's next major work, *Creative Evolution,* which first appeared in print in 1907 and in English translation in 1911. But here the rendering of the subject is less analytical, less psychological; in fact, time is viewed as an ethical-philosophical issue that transcends individual psychology, becoming instead identified with ideal progress and the transcendental meaning of evolution. The concern with the nature of consciousness and the relationship of the self as a conscious agent to the world of objects and matter persists, but it is less in terms of its personal meaning and more as a life principle, as a metaphysical creative force that gives meaning to the order of nature and to intelligence in its evolution from the lowest living organisms to the level of the human being.

In trying to define the meaning of existence, Bergson (1907) pointed out that the ego or the self as a conscious agent passes from state to state, changing without ceasing. But as what we do depends on what we are, we also are what we do. As Bergson put it: "We are creating ourselves continually" (p. 7).

Nevertheless, the key element to the meaning of existence is consciousness, for consciousness implies the awareness of a self as distinct from objects, a self within a world of distinct objects, objects that move or change in one or another aspect of their appearance. But for the idea of movement or change to occur, something must remain recognizable as such, the internal image of the object must remain the same, that image forming part of the self, the part of the self that remains unchanged, the self that endures. The feeling of duration is the consciousness of the self as the same within a changing state or image of oneself.

Movement and change occur all the time; but even though they recur in the same space, each occurrence is not perceived in the same way as the previous one; and that is true not only for different individuals, but even for the same one. For it is one's own duration rather than the duration of the external event that counts, what Bergson initially referred to as "real" or "concrete" duration, and later, writing about "creative evolution," designated as "lived" time. "Lived" time, like Kant's "intuition" of time, is not something "thought"; it is not a relation, but something concrete, like a feeling, a state of mind, an ego state, consciousness itself. To use Bergson's own illustration, the time one has to wait for a cube of sugar to dissolve in a glass of water is not proportional to the objective time it takes to complete the process, but to one's own patience or impatience on the occasion, to one's own particular affect state or state of consciousness. Time for Bergson (1907) is more than Plato's moving image of eternity; it is "a progressive growth of the absolute, and in the evolution of things a continual invention of forms ever new" (p. 364).

To a large extent, Bergson in his essay *Creative Evolution* dealt with time as a metaphor rather than experience, as an abstraction implied in the function of memory, or as a process implied in the fact of aging and decay. He wrote: "Wherever anything lives, there is, open somewhere, a register in which time is being inscribed" (p. 17). Bergson tried to deny that this is a metaphor by arguing that time is capable of "effective action," that time possesses "a reality of its own." But it is like the tradition of representing time as an old man with a scythe, or like saying that time is destructive as well as creative, which Bergson (1907) also said, directly or with the use of metaphors; for example, "Real duration is that duration which gnaws on things, and leaves on them the mark of its tooth" (p. 48).

Bergson returned once more to the subject of time late in his life with the essay *Duration and Simultaneity* (1922), in which he criticized Einstein's relativity theory. Bergson now defined the sense of time as the experience of a flow or passage which is self-sufficient, without reference to a passing thing or a flowing state of being, a sense of the self as transition that is experienced as duration. He compared it with the experience of a melody as one listens to it with eyes closed, allowing for the difference among the sounds to disappear, then losing track of the distinctive features of sound itself, except for the sense of an incessant transition, of something that precedes into something that follows, a sense of multiplicity without visibility and of succession without separation. Identifying this experience of inner transition with memory or consciousness, he concluded:

> It is impossible to distinguish between the duration, however short it may be, that separates two instants and a memory that connects them, because duration is essentially a continuation of what no longer exists into what does exist. This is real time perceived and lived. This is also any conceived time, because he cannot conceive a time without imagining it as perceived and lived. Duration therefore implies consciousness; and we place consciousness at the heart of things for the very reason that we credit them with a time that endures [p. 49].

The Self and the World of Internal Objects

Like Bergson, Heidegger came out of a philosophical tradition that was very familiar, if not congenial as well, to Freud and his early followers. Students of Heidegger and phenomenology, in general, point to Kant, Hegel, and Nietzsche as the precursors of the movement. Obviously, Husserl was very influential in Heidegger's thinking, even though Heidegger himself hardly acknowledged the connection.

Heidegger's major work, *Being and Time (Sein Undt Zeit)* appeared in the spring of 1927, the year after Freud published his *Inhibitions, Symptoms and Anxiety.* Its cornerstone concept is the self, the self in an object world *(Dasein).* Freud's use of the self in the form of "the I" *(das Ich)* is more ambiguous, especially in the term "ego" that his English translators adopted. But the use of the self by object relations theorists, beginning with Fairbairn, Winnicott, and Balint on the continent, and in this country Erikson, Jacobson, and Kernberg, is clearly related to Heidegger's concept.

Heidegger distinguished between the "authentic" self, referring to the awareness of one's own self as it really is or should be; and the "inauthentic" self, referring to the awareness of the self as others see it. One of the most important, if not the most important, characteristics of the "authentic" self and its existence in the world of objects is "care." Care is identified, at least partly, with concern (something that Winnicott [1958] also emphasized later in his own psychology of the "true" and "false" self). The idea of the *Dasein* (Being-in-the-world) is that the self is not merely a subject, but an integral part of a world of objects which are made for one another and cannot exist in a meaningful way without one another. The idea of objects as such is built into the essence of the self, being inherent in the structure of the self or mind.

Heidegger insisted on the necessity of defining object relations in terms of their meaning rather than according to objective characteristics or circumstances, and referred to his inquiry of existence as "hermeneutic," in the way that the French philosopher Paul Ricoeur (1970) and other friendly critics define psychoanalytic interpretations and theory: in terms of meaning rather than explanation.

Heidegger referred to the "inauthentic" or "false" self as "they-self," a self that exists, perceives the world, and operates in terms of the values or expectations of others. The "inauthentic" self is grounded on actuality, on what is real and at hand. Such an awareness defines a state of mind or mode of existence that is characterized by a sense of fatalism, inevitability, fallness, or "thrownness"—the sense of being thrown into this world rather than choosing to be born, our life being to a large extent determined by conditions and circumstances beyond our control. Such an awareness or state of mind is experienced as a mood that, depending on whether one likes one's circumstances or not, is perceived as depression, the perception of the actual as "fate;" or as elation, the perception of the actual as "fortunate" or lucky. To experience reality as actuality—in terms of the actual rather than the potential, as inevitable rather than open to possibilities—is to experience the self as object and the world of objects in an unconcerned, uncaring way, with curiosity perhaps but not with care. In contrast, to perceive the world as open to its influence, as a "potentiality-for-Being"—as the "authentic" self does—is to experience fear, or anxiety (dread).

Because man is concerned with his own security and tries to cope with what he has learned in the past, Heidegger feels that fear is determined by the sense of the past. The mood of anxiety or dread, on the other hand, which Heidegger described as very much like the experience of depersonalization and is allegedly rare, defines the function of self-awareness or understanding. It involves a mechanism of projection—projecting oneself into the dimension of "ecstasy" of the future. Heidegger used the Greek term "ecstasy" to indicate a dissociation of the self from the sense of time as such, to allow for the choice of perceiving oneself in terms of time's three dimensions: future, past, or present. Heidegger insisted that the self as part of the object's world *(Dasein)* is one with time, that man cannot be aware of his self without being aware of time in terms of one of its three dimensions. And in such a context, it is the identification or concern with one's future, including the awareness of one's existence as finite, that characterizes the true nature of the self. The identification of concern with the present or past, natural as it may be, is only derivative or defensive in nature, aiming at protecting the self from the awareness of its dreadful possibility of nonbeing, the awareness of death not just as the inevitable, objective, predetermined end of existence, but as something for which one is personally responsible.

According to Heidegger, we dread to be ourselves; dread "individualizes." To become conscious of one's consciousness is to feel alone and still care, to know that one is going to die while one is still alive; to be able to think of the unthinkable—that is dreadful.

Early Development of Time and Causality

Both Bergson and Heidegger influenced Piaget—Heidegger less clearly and belatedly, I suspect, during his years as a structuralist and epistemologist; Bergson rather early and, in spite of differences in their approach as researchers into the nature of the self, in particular as it concerns the relation of time and causality.

Bergson's concept of duration is James's (1890) "specious present" or "primary memory"—the overlap between fading and incoming impressions, the "germs" of past and future sense of time. Bergson's duration and James's specious present became Whitehead's (1927) perception of cause, the experience of one's self as a state of being that derives from a previous state, each experience characterized by "conformity to the antecedent." Echoing Bergson, Whitehead wrote: "Time is known to us as the succession of our acts of experience, and thence derivatively as the succession of events objectively perceived in those acts" (p. 35).

Echoing Whitehead is Piaget's (1937) sense of causality, "a sort of feeling of efficiency or of efficacy" (p. 227) that develops in connection with the child's early sensorimotor activity.

The realization that his action (e.g., reaching out for something or crying when hungry) is effective in getting him what he wants, observed Piaget (1937), creates in the infant a feeling of efficacy, generating along with it the consciousness of activity itself as a means of establishing a causative link between his desires and their fulfillment. Milk obtained is perceived as extending the act of sucking, the image of the nursing mother as extending that of the act of seeing. This early notion of causality, with its diffuse feeling of efficacy, is localized not in the self—which hardly exists at this stage—but in the focal point of the action, suspended in a space that is neither internal nor external in its reference, but at once internal and external. The systematic coordination of manual grasping (prehension) and sight within the context of the child's wishes invests objects with a certain "elementary permanence." Anticipating the ability to search for objects that have gone out of the child's perceptual field, this is the experience that things endure as a function of the child's action and desires—wishes that correspond to a wide array of needs other than visceral ones, notably the need for exploration. This early sense of permanence, built on the primitive notion of causality, is experienced as duration, and is the most elementary sense of time. As Piaget (1937) elaborated:

> this duration presupposes a representation of causality and not nearly a perception of it. Just as the permanence of the object is thus necessarily extended in representation of absent objects, so also the objectification and spatialization of physical causality sooner or later entail the representation of sequences not directly given in the sensory field [p. 294].

The perception of causes and effects marks the beginning of internal representation, with the feeling of efficacy ascribed to the self and experienced as self-esteem. According to Piaget (1937), the four building blocks of conceptual reality—namely, object (substance), space, causality, and time—evolve in an interconnected fashion and cannot be analyzed without reference to one another. The consciousness of time as a structure in the object world that entails seriation or succession—i.e., the notion of before and after, along with a more or less orderly sequence of events—begins to emerge as early as the first six weeks of life from a set of attitudes involving expectation, effort, and satisfaction or frustration of wishes and needs, which are experienced globally as a sense of duration. This elementary sense of time defines a primitive mental construction, one that is built solely on perceptions and personal displacements or object movements. In the absence of language or symbolic function, and without the benefit of internal representations, the infant's early cognitive reality is organized by means of a sensorimotor coordination of elementary functions and actions into spatiotemporal and causal mental structures.

Emerging during the initial two stages of sensorimotor intelligence, along with the acquisition of simple rhythms and habits or behavior patterns that take place in the first four or five months of life, early psychological duration is devoid of the awareness of sequence or seriation of events as external and independent of the self, simply because there exists no boundary as yet between personal activity and objects. The notion of space is there, but fragmented, allowing for no distinction between the internal world of experience and the external world of reality. During a third stage of cognitive development, characterized by a general unification of the notion of space and rudimentary differentiation of means from ends, the child begins to perceive the order of phenomena in relation to himself as their cause—a form of causality that Piaget defines as "magico-phenomenist." But once an object disappears from his perceptual field, the child shows only scant interest in it, being unable to experience duration in terms other than his own activity. However, as the process toward a system of permanent object-representations continues to unfold, and as the perception of coordinated positions and groups of displacements in space makes possible the projection of the notion of causality on objects, the notion of time as duration begins to acquire independence from one's own actions and, of course, desires. Beginning around the ninth month of life, such a development defines the fourth stage of sensorimotor intelligence, "the first one in which intelligence properly speaking appears" (Gruber and Voneche, 1977). During a fifth stage, which makes its appearance in the eleventh or twelfth month, the ability to objectify time becomes more general, extending from a few "privileged" events to the whole perceptual field. But it is still grounded in the present, without the power to evoke events in the past. During a sixth and final stage of the sensorimotor construction of reality that spans over the period of 18 to 24 months of life, the child becomes capable of evoking memories not linked to direct perception and thus gains the notion or awareness of the past as a meaningful point of reference, as part of his temporal perspective. As his behavior patterns indicate, he is now capable of planning his actions, ordering his displacements in a more or less systematic, insightful way.

Concerning the role of memory in the perception of time, Piaget (1937) has this to say:

> The child, from the first weeks of his life, knows how to recognize perceptual images. . . but this does not prove either that he knows how to evoke these images when he does not see them or that he is able, when he does see them, to localize in the past the memory of having seen them before [p. 326].

In other words, the development of the memory of recognition precedes that of the memory of localization or evocation. Unlike the adult experience of recognition, this early ability to remember does not involve a cognitive distinction between past and present, but only the ability to perceive the past as a "qualitative extension" into the present.

Clearly reflecting Bergson's (and Whitehead's) point of view, Piaget (1937) concludes:

> Primitive time is not time perceived without, but duration experienced in the course of the action itself. . . it is mingled with impressions of expectation and effort, with the very development of the act, experienced internally. [Primitive time is] simply the feeling of a development and of sequential directions immanent in the stages of consciousness [p. 326].

Conceptualization of Time

Piaget (1946) investigated further the development of the concept of time, showing how the acquisition of speech and the socialization of intelligence transforms the child's intuitive notion of duration into "operational" time, bringing the notions of duration and succession together into a coherent, logicodeductive system. The rudimentary notion of time, corresponding to an affective experience rather than to a cognitive appreciation of temporally arranged events—i.e., a sense of impatience that involves anticipation and implicitly the notion of future, or a sense of sadness that involves loss and the memory of the past—develops gradually into personal or psychological time, an experience endowed with continuity and universality. Duration can now be appreciated as a "continuous flux," lending itself to an approximate estimation of its length in terms of the particular life events it contains, an approximation that can be improved with exercise and the use of time-measuring devices—clocks and calendars—to become what Piaget (1946) describes variously as operational, rational, or physical time. He claims a "close relationship between psychological and physical time: both are coordinations of motions with different velocities, and both involve the same 'groupings.' This is only to be expected since both are derived from practical or sensorimotor time which, in its turn, is based on objective relations and on personal actions" (p. 277)

Piaget (1946), refers to the ability to conceptualize time as a cognitive function of the ego that matures with growth and experience. The term duration denotes both the subjective sense of the passage of time and the experience of physical duration as represented by age, a person's own and that of others. He examines empirically the development of the notion of time as implied in the process of aging, which he describes in terms of maturational stages equivalent to those observed in other areas of conceptual development, including that of physical time.

In embarking on his analysis of the child's conceptualization of time, Piaget (1946) pointedly dissociates himself from

Bergson's position regarding Einstein's theory of relativity, stating:

> Bergson's entire philosophy, as well as the large number of purely psychological works influenced by it, have stressed the great importance of the concepts of inner duration and of psychological time. Yet far from seizing on the possible convergence of Einsteinian time and lived duration, Bergson himself tried to contrast them in a little work [*Duration and Simultaneity*] that caused quite a stir at the time [p. 26].
>
> . . . Bergson, far from applauding the fact that Einsteinian time presents physics with a much closer model of psychological time (we might say of Bergsonian time) than Newtonian time did, challenged relativity theory with the claim that relative time was a characteristic of life alone [p. 278].

Indeed, as an empirical scientist, Piaget found much to criticize or reject in Bergson's ideas of time. Concerning the self, however, conceived as an experiential structure that is invested with the consciousness of time, he could only agree that "lived durations are not simply intervals but, as Bergson so rightly put it, the 'very stuff of reality'" [p. 276].

III

EARLY DEVELOPMENT OF TIME: AN OBJECT-RELATIONS POINT OF VIEW

THE NOTION OF TIME IS the result of the integration of two general areas of psychological development: rudimentary ego functions allowing for the perception of space, i.e., the grasp of spatial relations such as movement, simultaneity, and succession; and primitive motivational states conceptualized as internal objects. To begin with, time as a psychological entity is predicated on the development of consciousness: the awareness of the self as an internal image or representation separate from object images or representations. But for the consciousness of time to grow, it is necessary to acquire the consciousness of movement, i.e., the awareness that objects about oneself move or change; and the consciousness of objects as unique, continuous, relatively stable.

Temporal Cycles and Need Fulfillment

One of the first psychoanalytic writers to deal with the subject in some detail, Harnick (as quoted by Schilder, 1942, and by Doob, 1971) suggested that the awareness of the flow of time has its origin in anal eroticism—the frequency of defecation, its intervals, the length of the process, how long it may be postponed successfully, and so on. But he also implicated the experience of hunger, speculating that time in the unconscious stands for the introjected, "devoured" father who has become fecal matter.

While agreeing that anal-erotic experiences are responsible for the development of schedules that help the ego to master reality, Fenichel (1945) pointed out that internal biological rhythms, such as breathing and pulse, convey kinesthetic sensations that are even more basic in helping the child to differentiate time intervals and to acquire the concept of time in general. In discussing the notion of time and its relation to the sense of reality and purpose, Oberndorf (1941) reiterated the view that anal experiences are mainly responsible for the development of schedules that help the ego to master reality.

The idea that the infant acquires the meaning of time in the intervals between feedings was advanced first by Hanns Sachs (as quoted by Bergler and Róheim, 1946). Yates (1935) wrote about patterns of body rhythms that, during the oral stage of psychosexual development, become associated with the periodic frustration and satisfaction of the infant's needs, especially the need to be fed, thereby providing the basis for the development of the sense of time. Likewise, Meerloo (1954) assumed that the sense of "time and expectation" develops out of the infantile experience of hunger. A related but broader view was advanced by Erikson (1956), who wrote: "The experience of temporal cycles and of time qualities are inherent in and develop from the initial problems of mounting need tension, of delay of satisfaction, and final unification with the satisfying 'object'" (p. 97).

In a similar vein, Gifford (1960) suggested that the emergence of the ego as an organized structure and the acquisition of the sense of time come about as the infant learns to adapt his pattern of sleep-wakefulness to the diurnal periodicity of maternal attention to his needs. As Gifford put it: "This early adaptation to the outer world is transmitted to the infant as a function of time" (p. 26).

Anticipation and Anxiety

Freud (1914, 1920, 1933) ascertained that the concept of time and its genesis are inherent in the organization and function of the ego. Hartmann, Kris, and Loewenstein (1946) pointed out that the process of maturation entails the ability to postpone immediate gratification and the development of the function of anticipation:

> Maturational changes proceeding during the second half of the first year give the child further control of his own body and enable him partly to master the inanimate objects in his life space. Some kind of anticipation of future events plays its part in each of these operations. They represent a central function of the ego; that which makes the transfer from the pleasure principle to the reality principle possible [p. 22].

Later on, referring to the ego functions of motility and perception, Hartmann (1964) made the following suggestions regarding the development of the ability to experience time:

> Many aspects of the ego can be described as detour wishes; they promote a more specific and safer form of adjustment by introducing a factor of growing independence from the immediate impact of present stimuli. In this trend toward what we may call internalization is also included the danger signal, besides other functions that can be described as being in the nature of anticipation [p. 115].

At this point, Hartmann simply referred his readers to Freud's thoughts about the relation of the ego to time perception, about which Anna Freud (1965) had this to say: "[It is] from primary to secondary functioning that we get anticipation of the future, between wish and action directed toward fulfillment" (p. 92).

The idea of an internalized danger signal and the mental function of anticipation as precursors of the time sense finds ample correspondence in the psychiatric and psychoanalytic literature. Arieti (1947) explicitly attributed the evolution of anxiety and of symbolic processes to expectancy and anticipation. He defined expectancy as "the capacity of the subject to anticipate certain events while a certain external stimulus is present," and anticipation as "the capacity to foresee or predict future events, even when there are not external stimuli which are directly or indirectly related to those events" (p. 471). He conceived of these two mental processes as being ontogenetically contiguous, the latter growing out of the former. Phylogenetically also, according to Arieti, the process of anticipation is antedated by that of expectancy, which may be identified in subhuman, even very primitive species, while the process of anticipation is to be found only in man. Spitz (1972), on the other hand, has shown that anticipation as a psychological process evolves out of the neonate's conditioned reflex activity, which "forms the first bridge between somatic neural physiology and psychic activity" (p. 727). Rapaport (1945) pointed out that the notion of anticipation is intimately related to time experience, in particular to time as perspective, and Hartmann, Kris, and Loewenstein (1946), that "anxiety as a signal can operate only when the child has learned to anticipate the future" (p. 28). After reviewing the literature in natural sciences as well as in philosophy and history, Fraser (1972) concluded: "anxiety and the sense of time are mutually generated modalities of the mind" (p. 119).

Primitive Objects and Affective Dispositions

The view that time and affects interdigitate has been explicitly or implicitly held by psychoanalysts and academic psychologists alike. Thus, according to Eissler (1955):

> The interplay of emotions, their speed, their gradients, the ego's acceptance, objection, or denial of them and many more factors which are involved in the appearance and disappearance of emotions, appear to leave their imprint upon the way time is experienced [p. 281].

More specifically, Spitz (1972) has indicated that affect is the element responsible for the acquisition of the notion of time as duration, which in turn makes the function of perception meaningful. In his words: "The percept can only acquire existence after affect has endowed it with duration, with biological time" (p. 733).

The development of the sense of time seems to determine and in turn to be determined by the transformation of primitive affective dispositions (defined broadly as pleasure and unpleasure) into specific affects like anxiety, depression, boredom, elation, and so on. Unfolding with the maturation of the ego apparatus in its perceptive-cognitive functions, the self's awareness of the continuum of time is predicated on the perception of its efficacy regarding some adverse or challenging reality, within or without. It is the inefficiency of the self in terms of experienced tension or unpleasure that cannot be mastered immediately, along with a growing awareness of the possibility that relief from the "good" object will soon come, that gradually establishes the sense of time, first as the notion of duration or protensity (experienced as the passive process of waiting in anticipation), and then as temporal perspective in its rudimentary form of intentionality (experienced as the active process of expectation). Such a formulation is consistent with Piaget's (1937) understanding of the roles of anticipation and expectation in the development of the notion of causality, which, along with the notion of space (encompassing the notions of velocity, simultaneity, and succession), constitute in his theory the necessary psychophysical determinants of time.

According to psychoanalytic object-relations theory, the capacity to anticipate need fulfillment protects the infant from the traumatic experience of helplessness that frustration of urgent needs and flooding of the ego with excitation may create. The capacity to anticipate need fulfillment is giving rise to a putative hallucinatory experience that involves the memory of the "good" breast or "good" mother—an experience that, according to Freud (1925b), marks the beginning of the thinking process and, according to Erikson (1956), "contributes temporal elements to the formation of basic trust" (p. 97). As tension rises and the mother is not yet there, the "good" object image or representation emerges protectively in fantasy and unites with the self-image in a need-fulfilling hallucinatory experience; but if the mother's arrival is further delayed, it begins to fade away rapidly. As the infant tries to hold on to it and unpleasure increases, the uncertain "good" object image begins to turn into a "bad" one. It is the effort to hold on to the "good" object image and to expel the "bad" one that presumably creates the ability to anticipate the future; and while tension is still tolerable, the situation is experienced as anxiety. As tension continues to rise, however, the "bad" object image tends to prevail. The "good" image fades away and anxiety turns into fear and anger, "impotent rage . . . in which anticipation is obliterated" (Erikson, 1956, p. 97). If the mother's arrival is delayed further, the infant's brittle ego apparatus becomes flooded with excitation, his mental state regressing to the level of an undifferentiated "bad" self-object image—all of which amounts to an experience of catastrophic dimensions. Such an experience would be timeless.

Eventually, and as object constancy develops, the early hallucinatory process is replaced by the ability to anticipate need fulfillment, i.e., confidence in the probability that the "good" object will arrive, once more emerging out of the fused internal representation that combines the "good" and "bad" object image. As this fused "good" and "bad" object image becomes consolidated into a stable, realistic object apart from the self, the trustful, rather passive ability to anticipate a "good" outcome is employed in the self's cathexis of the environment, which is projectively experienced as continuous—i.e., as possessing the attribute of time as duration. At the same time the anxious, rather effortful ability to anticipate a "good" outcome is employed in the projection of the self-representation beyond the confines of the immediate environment, the whole situation acquiring the quality of time as perspective and, more specifically, as future.[1]

In the specious present of the infant, tension or need is presumably experienced as rising and falling, as growing in urgency and then, if satisfied, as dying away rather than as merely experienced and then not experienced. Yet a need does not have to be satisfied in order to be experienced as receding, for there is a

[1] The self-image or representation is a psychological structure made up of the ideal self, the ideal object, and the actual self-image accompanied by a feeling state that defines a person's self-esteem. It is an experiential entity predicated on the concept of consciousness. Ego, on the other hand, is a system of functions that, along with the superego and the id, belongs to the larger structure of the psychic apparatus. It is a conceptual entity on a higher abstraction level than the system of the self, and describes relationships as manifested in defensive operations and the process of adaptation. As such it contains the system or structure of internal object relations. Even though it can be experienced in its manifestations, it cannot experience, except insofar as the self can. In other words, the ego is the seat of experiencing only insofar as experiencing takes place in the self. A balanced self-system is experienced as self-esteem.

built-in rhythm to the excitability of the organism, regardless whether the stimulation is there or not. When the tension becomes too persistent and relief is not forthcoming, the specious present (or experience of duration) tends to become shorter, the number of discernible gradations within it fewer, the perception of time's flow slowing down and then stopping; in other words, the perception of time as duration disappears along with the perception of change. The same would obtain if the need were met with satisfaction. Once satiated, the infant ceases to perceive change, his perception of time as flowing by slows down, and he falls asleep or passes into a state of pleasant timelessness.

Speaking of need and tension generically may suggest alimentary functions—as indeed a number of psychoanalytic writings do when referring to the origins of time. Yet it is known that after the first days of postnatal life the infant may remain awake and alert without being necessarily hungry or in any perceptible way uncomfortable. And in this state of alertness, he can perceive movement or, in general, change in his environment other than the coming and going of his mother. Apparently ontogenetically determined, there is in the newborn child something like a behavioral preference to contour and movement, an unlearned orientation to change that, according to Kagan (1970), "is clearly adaptive, for the focus of change is likely to contain the most information about the presence of his mother or of danger" (p. 828).

That such an orientation to movement or contour might alone account for the development of the concept of time seems, however, unlikely. Object-relations theory assumes that for the infant to acquire the notion of time, he should go through a motivational process involving his mother as a need-fulfilling as well as frustrating object. And Piaget (1966), who has tried to establish a psychophysical parallelism regarding the concept of time, holds that the experience of duration develops as a function of a wishful state of waiting:

> It appears that, psychologically, the time which impresses the child the most at the outset is the time of the events

> which surround him. The child's attention is centered more on things going on around him than in his own introspection. Yet sooner or later the child will experience waiting for a longer or shorter time for the fulfillment of a desire [p. 211].

Fraisse (1963), whose experimental work on time has been heavily influenced by Piaget, also states: "the young child has intuitions not only of speed and distance but also of duration. He sees the latter in the elementary form of an interval which stands between him and the fulfillment of his desires" (p. 277).

Development of Object Constancy

Acknowledging the importance of "felt anxiety" as an experience that develops along with the ego's capacity to anticipate and predict, Benjamin (1961) described a contributory and perhaps necessary condition for the development of time sense: the "maturational organization of aggression as such into object-directed hostility and anger, with the resultant marked increase in fear of object loss" (p. 662). Besides the implication of aggression and related affects in the emergence of time as perspective, Benjamin's point underscores the significance of object constancy. Jacobson (1964) also has pointed out that it is the development of object constancy that determines the concept of the self as an entity with continuity and direction, making it capable of experiencing time and specific affects.

Rudiments of the sense of time should begin to develop along with the process of differentiation of self- from object representations. On the basis of the available literature of psychoanalytic object-relations theory, Kernberg (1976) dates this development as far back as the third or fourth month of life. But the establishment of the notion of time as duration and tense or perspective has to wait until the second half of the first year of life, when the existence of object constancy becomes evident. According to Spitz (1957), such evidence may be inferred in the phenomenon of "stranger anxiety" observable at the age of eight months. Others, following Piaget's criteria for object permanence, do not find such evidence until sometime in the second year. According to Piaget (1937), object permanence is based on new discoveries the infant makes concerning space, time, and causality in relation to a desired object that is no longer in view, reachable or perceivable. Citing the work of his fellow countryman and psychoanalyst Charles Odier, Piaget (1962) elaborated that "this schema of the permanent object is caused by feelings and object-relations. In other words, the affective relations of the child with the mother-object, or other persons, are responsible for the formation of a cognitive structure" (p. 129). But Piaget's concept of object permanence, which the child acquires between the twelfth and eighteenth month of life, may not be equivalent to the psychoanalytic concept of object constancy (Fraiberg, 1969), which establishes itself affectively rather than cognitively at a somewhat earlier age, before the acquisition of a generalized understanding of time.

Mastery of Temporal Perspective

In considering the development of temporal perspective, the importance of memory as it represents past experience is obvious. James (1890) and Bergson (1907) have written at length about it. Russell (1948) put it as follows:

> There are two sources for our knowledge of time. One is perception of succession within our specious present, the other is memory. Remembering can be perceived, and is found to have a quality of greater or less remoteness . . . [but] our knowledge of the past . . . depends upon some postulate which is not to be discovered by mere analysis of our present rememberings [p. 216].

Memory, whether conceived as conditioning, as long-term information storage, or as mental representations and internal object relations, is not necessarily accompanied by the sense of "pastness." A stored sensory experience is not equivalent to the memory of something that has happened, something in the past. For memory to acquire the quality of "pastness" it is necessary to perceive the fact that the recalled experience can no longer be altered, either by the subject's own action, or passively by some intervention from the outside. For the infant to have any sense of time, he should find himself in a state of relative need-tension. So long as he is able to maintain a trustful expectancy of mother's presence while she is absent, he will have the experience of time as duration—the equivalent of the adult normal experience of time passing, flowing by. But unlike the adult experience of duration, which normally "seems most keyed to remembrance of things past—to retrospection" (Ornstein, 1969, p. 22), in the infant's specious present, memory as long-term storage plays no part, except insofar as it may potentiate a hallucinatory experience. The hallucination of a satisfactory experience—the mother's need-fulfilling presence—activated by a current need, is conceivable only as an event that evokes the sense of present, not past. The moment that the need-tension rises again, signifying that the hallucinatory experience is becoming impotent, the self begins to feel threatened. Anticipating the worst, helplessness and disintegration, a trauma, the infant now experiences anxiety—a "basic affective response," as Joffe and Sandler (1965) put it, containing a rudimentary sense of future.

As Piaget (1946) has pointed out, "memory recalls or anticipates" (p. 2). And paradoxical as it may seem, anticipation must occur in memory before recall. A memory is localized in time and space, but it is first experienced as anticipation, as future satisfaction or frustration. As Fraisse (1963) put it: "In all . . . initial instances of temporally organized behavior it is the past which gives meaning to the stimulus, turning it into a signal, but the signal gives rise to behavior which is oriented toward the future" (p. 154).

If indeed the sense of time as perspective begins as the ego becomes able to differentiate between present need and future fulfillment—in other words, if the sense of future develops before that of past, as a number of authors (Stern, 1924; Schur, 1953; Whitrow, 1961; Fraser, 1975), besides Fraisse, maintain—then at what point does the sense of past develop? When does the past enter the child's awareness of time as a dimension of experience?

In discussing the growth of the mind in terms of object-relations theory, Kernberg (1976) assumed the following sequence of events: At a certain point, the infant comes to realize that the self as a subject of gratifying affective experiences with the mother is the same as the self interacting with mother under the influence of frustrating, anxiety- or rage-producing experiences. Presumably out of the integration of these opposite "bad" and "good" self-representations and, more specifically, at the junction of the development of the self-concept as an independent reality in the ego, there develops the awareness or acknowledgement of the past as a dimension of experience. The infant is now able to accept "bad" past rather than deny it, to acknowledge a narcissistic hurt as a fact, unavoidable but not catastrophic, a mitigated less-than-total defeat.

The whole "schema," to use Piaget's term for a cognitive-affective structure,[2] constitutes a mental state typical for the age — a developmental phase that Melanie Klein (1935) first identified as the "depressive position." Winnicott's (1958b) observation that such a developmental phase makes possible the affective reactions of guilt and concern may be elaborated as follows: The infant feels guilt to the extent that he feels his aggression has happened and cannot be renounced; and he feels concern for the object that has been, in the infant's fantasy, hurt by the aggression but remains "good," i.e., present and loving rather than getting lost or hateful.

According to object-relations theory, the ego does not emerge as a separate, effective structure apart from the id until the mechanism of repression gets established, somewhere in the second year of life (Kernberg, 1976). As the ego and the id become differentiated from their common matrix, repression establishes itself as the main defensive operation of the ego apparatus, replacing primitive, psychoticlike mechanisms such as splitting, projective identification, denial, isolation. It would be reasonable to assume that this is the point at which the sense of time and its differentiation into future and past become consoli-

[2] More accurately, a "schema" is "a representation of experience that preserves the temporal and spatial relations of the original event, without being necessarily isomorphic with the event" (Kagan, 1970, p. 827).

dated, appearing in the child's social awareness and communication through language. Evidence from experimental literature seems to support such an assumption. Thus, in studying the verbal understanding of time concepts in young children, Ames (1946) found that "time" words and their use become meaningful in a fairly uniform sequence, which is approximately the same for all children. Her subjects were able to respond to "soon" as a time concept connoting the future by waiting at 18 months, and to use it in a phrase at 24 months. However, they were not able to verbalize spontaneously the word "tomorrow" until 30 months, and only after they had started using the word "today." Mastery of the word "yesterday" came last, at 36 months.

Even though the understanding and use of words connoting time provides a fairly reliable means of assessing the child's ability to orient himself temporally, it is by no means the earliest indication of his ability in this regard. As Werner (1957), and Werner and Kaplan (1963) have pointed out, the very young child expresses his temporal values in terms of affective language and goal-oriented behavior, a temporal reference being implicit rather than identified as such in his words. At first, a temporal perspective is implied in purely inferential one-word ("monoremic") expressions fused with particular events and affects; then, with experience, temporal meaning is more and more carried by action words (verbs), now increasingly differentiated from thing-words (nouns). The mastery of temporality in terms of future, present, and past is eventually facilitated by the learning of adverbs and other words that indicate tense.

Later Stages of Development

In contrast to the ego—where rudiments of time exist as the forerunners of basic functions such as perception, motility, and memory—the id or dynamic unconscious starts out absolutely timeless (Freud, 1901, 1915, 1933; Bonaparte, 1940). In terms of object-relations theory,the original timelessness of the id corresponds to an early stage of mental development during which the frustrating mother is experienced as absolutely "bad" or as a "bad" fused self-mother representation. But, gradually, a sense of time builds up in the id, too, as successive layers of dissociated and repressed object relations establish a history. For defensive purposes, however, the unconscious eliminates or distorts time —as can be seen in dream work and in borderline or psychotic conditions—so that the eventual timelessness of the unconscious is but a regression to the level of its original absolute timelessness.

An important determinant of time is to be found in the integration of self- and object representations, whereby complete independent perceptions become identified with the same object—in other words, when an object can be experienced in different ways at different times and still remains the same. Such an accomplishment becomes a measure of the ego's strength and reflects its ability to deal comfortably with the dimension of time, making it possible to explore the consequences of one's acts, real or imaginary, and thereby to arrive at a prediction of change on the basis of one's own effort. Carried out first by the ego, this ability later becomes a superego function. To the extent that inner rewards or punishment through guilt are functions of the superego, its maturation powerfully reinforces the sense of time.

> Insofar as the superego is the agency of inner standards, demands, ideals, hopes and concerns in regard to the ego, the agency of inner rewards and punishments in respect to which the ego experiences contentment or guilt, the superego functions from the viewpoint of a future ego, from the standpoint of the ego's future which is to be reached, is being reached, is being failed or abandoned by the ego. Parental and other authorities, as internalized in the agency of the superego, are related to the child as representatives of a future and of demands, hopes, misgivings or despair, which pertain to an envisaged future of the child. The superego watches, commands, threatens, punishes, forewarns, admonishes, and rewards the ego, it loves and hates the ego. All this we can do with ourselves only insofar as we are ahead of ourselves, looking back at ourselves from a point of reference which is provided by the potentialities we en-

> visage for ourselves or of which we despair. Conscience speaks to us from the viewpoint of an inner future, whether it tells us what we should do or how we should behave in the future, or whether it judges past and present deeds, thoughts, and feelings [Loewald, 1962, p. 265].

The observations of early psychoanalytic workers with regard to the influence of the anal phase of psychosexual development on the formation of the concept of time should still be valid insofar as the later evolution of the concept is concerned, in particular its relation to the superego and its involvement in neurotic symptoms, characterological traits, and sociocultural styles. In the process of toilet training, the toddler identifies with the preoedipal mother who controls time, the wish and ability to control time reflecting the wish and ability to be in control of the need-tension child-mother relationship in its internal representation. As Colarusso (1979) has pointed out, with the child's increasing ego development, the wish to control time loses its imperative, instinctual quality, becoming gradually an autonomous ego function. Such derivative affects and structures of the oedipal phase as the fear of death, incest taboo, and the attraction toward an unattainable object that characterizes romantic love are inherently related to the experience of time and its conceptualization as fate or as progress, as the inevitable or as the eternal (Panel, 1972). Eventually, as the child's mind unfolds in its maturational schedule, the emergence of the ability for abstract operations will make it possible to deal with time as a measurable entity, as a standard of reference in terms of clocks and calendars, as a universal rather than simply personal, self-centered reality.

IV
AFFECTS AND THE SELF

More than twenty-five centuries old, Sappho's elegant fragment of a poem conveys a vivid sense of loneliness and longing, which would not be possible without the consciousness of time:

> The Moon and Pleiades
> Are Set. Night is half
> Gone and time speeds by.
> I lie in bed alone.

And again no sense of time would be possible without the consciousness of self apart from and longing for another, the self and the absent other. For there is no meaningful concept of the self without consciousness, and there is no consciousness without self—the self is the container of consciousness. Affects such as loneliness and longing develop after the emergence of the self and are by definition conscious.

The Purposeful Nature of Affects

According to classical psychoanalytic theory, affect is the experience of some kind of inner discharge—hormonal, neurovascular, or neuromuscular (Rapaport, 1953). The idea was prevalent at the turn of the century, when Freud was trying to elucidate the dynamics of hysteria. As Bergson (1889), one of the most influential teachers of the time, put it:

> . . . eliminate . . . all trace of organic disturbance, all tendency towards muscular contraction, and all that will be left of anger will be the idea, or, if you still insist on making it an emotion, you will be unable to assign it any intensity [p. 30].

Citing Spencer's statement that "fear, when strong, expresses itself in cries, in efforts to escape, in palpitations, in tremblings," Bergson commented:

> We go further, to maintain that these movments form part of the terror itself: by their means the terror becomes an emotion capable of passing through different degrees of intensity. Suppress them entirely, and the more or less intense state of terror will be succeeded by an idea of terror, the wholly intellectual representation of a danger which it concerns us to avoid. There are also high degrees of joy and sorrow, of desire, aversion and even shame, the height of which will be found to be nothing but the reflex movements begun by the organism and perceived by consciousness. . . . The acuteness of these emotions is estimated by the number and nature of the peripheral sensations which accompany them. Little by little, and in proportion as the emotional state loses its violence and gains in depth, the peripheral sensations will give place to inner states; it will be no longer our outward movements but our ideas, our memories, our states of consciousness of every description, which will turn in larger or smaller numbers in a definite direction [p. 30].

Bergson's view reflects Darwin's and James's ideas about the nature of emotion. James (1884) dealt first with the subject in an article entitled "What is an Emotion?" And when barely a year later the Danish physiologist, Lange, published a similar view, the two became known as the James-Lange theory of emotion. In essence, the theory viewed affect as the experiential consequence of peripheral, visceral or vasomotor changes in the body brought about by the perception of relevant external stimuli. Familiar, even though not satisfied, with these views, Freud (1915) postulated further that if the cognitive aspect of an affect becomes repressed, in other words if the ideational or content of the affective or emotional process is defensively eliminated from consciousness, the experience loses its specificity and is no longer subjective and affect-laden. Affective experience is thus converted into a symptom that carries a different experiential quality, which may give the impression of indifference (as in conversion hysteria), or of free-floating anxiety (as in anxiety states). Freud also suggested that the organic disturbance or inner discharge could be displaced to an idea related but removed from the situation he defined in terms of conflict. Neurophysiologist Walter Cannon (1915) proposed his "emergency theory," according to which affects constitute signals for action. The arousal of affect serves the biologic purpose of preparing an organism to take action — flight in response to fear, fight in response to rage. But the purpose of preparing oneself requires implicitly temporary elimination of action. The preparation period over, the ensuing action — fight or flight, approach or avoidance — would as well eliminate the affect.

Rejecting the James-Lange position, Cannon (1929) maintained the affect, or "felt emotion," rather than arising from somatic, visceral, or vasomotor changes, originates from changes in the brain itself, specifically from discharges in the thalamic area, where it is elicited by cognitive states (cortical processes), which otherwise and in spite of any peripheral changes would remain "pale, colorless and destitute of emotional warmth" (p. 370).

In merely accounting for the quality of the emotional experience, Cannon's view is not so different from the James-Lange theory. According to both views, the experiential or "sensory" aspect of the affect is due to internal discharges: a peripheral or, rather, centripetal (visceral or thalamic) disturbance elicited by a central or centrifugal (cortical) process. The fundamental difference in Cannon's theory is in viewing affects as purposeful. Likewise, in his second theory of anxiety, Freud (1926) emphasized the adaptive nature of affects — at least as far as the affect of anxiety is concerned. What arouses anxiety is the ego's perception of the self in danger and the need to do something about it. What prepares the organism for defense or action involves a process of cognition — the awareness of internal changes. This process of cognition usually involves more than a perception or a memory; unless it is felt as merely pleasant or unpleasant, it involves an interpretive, biologically purposeful defensive operation.

The functional significance of affects, noted first by Darwin (1872) and subsequently elaborated by Cannon and Freud, was emphasized by Bergson (1889) in connection with his discussion of time as a dimension of the self, and by implication of affect. Thus he wrote:

> It seems indeed somewhat improbable that nature, so profoundly utilitarian, should have here assigned to consciousness the merely scientific task of informing us about the past or the present, which no longer depend upon us. It must be noticed in addition that we rise by imperceptible stages from automatic to free movements, and that the latter differ from the former principally in introducing an affective sensation between the external action which occasions them and the volitional reaction which ensues. Indeed, all our actions might have been automatic, and we can surmise that there are many organized beings in whose case an external stimulus causes a definite reaction without calling up consciousness as an intermediate agent. If pleasure and pain make their appearance in certain privileged beings, it is probably to call forth a resistance to the automatic reaction which would have taken place: either sensation has nothing to do, or it is nascent freedom. But how would it enable us to resist the reaction which is in preparation if it did not acquaint us with the nature of the latter by some definite sign? And what can this sign be except the sketching and, as it were, the prefiguring of the future automatic movements in the very midst of the sensation which is being experienced? The affective state must then correspond not merely to the physical disturbances, movements or phenomena which have taken place, but also, and especially, to those which are in preparation, those which are getting ready to be [p. 33].

Bergson also suggested that consciousness establishes itself in the process, the emergence of consciousness and of the self being directly related to the experience of affect as a purposeful signal elicited by conflict and in particular by the awareness of danger, i.e., the awareness of the possibility of injury to the self in the future and thus the awareness of the future, which is at the basis of the affect of anxiety.

What makes an affect subject to consciousness is the spatialization of time, the development of the sense of time as perspective, future or past, out of the sense of duration, the sense of presence or absence. What creates the sense of duration is the experience of unpleasure in association with the recognition of the absence of a gratifying object, a prolonged enough experience of tension that is relieved by the appearance of a gratifying object—mother's presence.

The experience of tension, identified with the image of a gratifying object in its absence, creates the sense of expectancy, which eventually becomes the sense of future. The experience of tension in the context of expectancy becomes the affect of anxiety, the prototype of all meaningful affects.

Affect and Cognition

Affect or emotion is a psychophysiological process with behavioral components that refer to the expression of whatever underlies the process. An important aspect or component of this process is its subjective experience, which in most cases is enough to define a particular affect. But what if this subjective component is not there—which is often reported in clinical practice; is there still an affect? Are there unconscious affects as there are unconscious ideas?

The notion that an affect can be unconscious creates problems when it is defined in terms of its subjective, experiential value. Freud (1915) circumvented the issue by referring to "unconscious affects" as "possibilities" or affective dispositions, rather than as actual psychological processes, a status only unconscious ideas qualified for, in his opinion:

> . . . unconcious ideas continue to exist after repression as actual structures in the system *Ucs.*, whereas all that corresponds in that system to unconscious affects is a potential beginning which is prevented from developing. Strictly speaking, then, and although no fault can be found with the linguistic usage, there are no unconscious affects as there are unconscious ideas. But there may very well be in the system *Ucs.*, affective structures which, like others, become conscious. The whole difference arises from the fact that

> ideas are cathexes—basically of memory-traces—whilst affects and emotions correspond to processes of discharge, the final manifestations of which are perceived as feelings [p. 178].

And referring to anxiety, the prototype of affects for psychoanalytic theory, Freud (1930) wrote:

> Anxiety is always present somewhere or other behind every symptom; but at one time it takes noisy possession of the whole of consciousness, while at another it conceals itself so completely that we are obliged to speak of unconscious anxiety or, if we want to have a clearer psychological conscience, since anxiety is in the first instance simply a feeling, of possibilities of anxiety [p. 135].

Bleuler's (1911) reference to inhibited affects as affects *in nascendi* is reminiscent of Freud's conceptualization of "unconscious affects" as mere affective dispositions. But while theoretically psychoanalysts might not disagree with Freud and Bleuler, in clinical practice most of them find it hard to avoid using the concept of the unconscious not only with ideas, but with affects as well (Pulver, 1971).

A related problem is the border with cognition, the fact that the distinction between affective and cognitive process is not as clear as Aristotle's (384–322 B.C.) or, for that matter, Freud's categorization of affect as distinct from cognition would have us believe. It is not only that cognition plays an important role in the determination of particular affects, but mental states as such are both affective and cognitive at the same time. As Dewey (1886), James's contemporary, pointed out:

> [Affect] signifies not a special class of psychical facts, like memory or conception, but *one side of all mental phenomena.* It is not a particular group of psychical experiences, occurring now and then in our mental life; it is coextensive with mental life; it is its *internal* aspect [p. 246].

Or, as McDougall (1923), whose instinct psychology competed for a time with that of Freud's, more globally put it:

> We often speak of an intellectual or cognitive activity; or of an act of willing or of resolving, choosing, striving, purposing; or again of a state of feeling. But it is generally admitted that all mental activity has these three aspects, cognitive, conative, and affective; and when we apply one of these three adjectives to any phase of mental process, we mean merely that the aspect named is the most prominent of the three at that moment. Each cycle of activity has this triple aspect; though each tends to pass through these phases in which cognition, conation, and affection are in turn most prominent; as when the naturalist, catching sight of a specimen, recognizes it, captures it, and gloats over its capture [p. 266].

Actually, affects may be viewed as ego states or structures of the mind that, as McDougall pointed out and Freud (1915) himself acknowledged, involve not only affectivity and cognition, but also conation, i.e., desire (appetite) and will. As Piaget (1964) said about "personal schemata"—an equivalent concept—they are "both intellectual and affective. We do not love without seeking to understand, and we do not even hate without a subtle use of judgment" [p. 207].

Coping and the Awareness of Novelty

According to the James-Lange theory, academic psychology and, indeed, classical psychoanalytic psychology, there is a sharp distinction to be made between perception and affect. These traditional views hold that perception—"a more intuitive type of cognition" (Brunswik, 1957, p. 7)—is the central elaboration of sensory (peripheral) processes, while affects or emotions are the elaboration of neurohumoral mechanisms and the experience that results from the perception of internal changes or discharges. But recent developments in neuropsychology and allied fields have made such a distinction appear questionable. An impressive body of new experimental findings (Pribram, 1967, 1970, 1971; Leeper, 1970; Schachter, 1970) suggests that affects are not simply the felt, conscious aspects of internal discharges but, rather, the experiential equivalents of an organismic state of disequilibrium that occurs when a person is confronted with a novel situation. What defines novelty in the case is a discrepancy between the presenting stimulus and a cortical image of action, the absence of fitting coping devices or defenses—strategies that are encoded in ontogenetic (and probably phylogenetic as well) memories, what psychoanalysts conceptualize as "psychic structures" (Rapaport, 1953), neurophysiologists as "neuronal models" (Sokolov, 1960), and system theorists as "programs" or "plans" (Pribram, 1967). Affects signal changes in the homeostatic organization of the brain, changes in response to fluctuations of stimulus input from the external or internal environment. Whereas the perception of the outside world registers an inner, central image that is attributed to objects and is felt as the consciousness of something out there, the perception of stimulation from within (neurohumoral changes, instinctual drives, superego demands) produces something of an image that is attributed to the self—a self-image in a state of arousal, excitement, bewilderment, uncertainty.

Consciousness implies a break in the continuity of one's perception of the world, the awareness of a contrast as such rather than a contrasting relationship, contrast that is integrated into the perceptual field. As Whitehead (1929), anticipating modern neuropsychology, put it: "Consciousness is the feeling of negation. . . . It finally rises to the peak of free imagination, in which the conceptual novelties search through a universe in which they are not datively exemplified" (p. 245).

Such a discontinuity in the perception of one's objective environment—whereas a figure stands out of the ground or the ground is perceived as surrounding a figure—results in the awareness of objects with qualities of their own: shape, color, intensity. In an analogous way, self-awareness implies a break in the continuity of one's perception of one's inner environment—the awareness of self- as separate from object representations. Affects define the quality of self-awareness, they are properties of self-consciousness.

As we saw in the previous chapter, self-consciousness is born out of tension arising from internal object representations that threaten the integrity of the self, the boundaries that differentiate the image of the self from that of an object image. Self-consciousness, however, or the mere awareness of an object as separate from the self, is not enough to elicit the experience of a specific affect; it may give rise to a motivational state that is experienced as tension, but tension as appetite (desire) or aversion rather than as affect *per se.* For an affect to occur there should arise a discontinuity in one's self-awareness, a break in the putative boundaries of the self.

Essentially the phenomenon of affect signifies an attempt to deal with a state of mental disequilibrium, not in terms of action but by the regulation of perceptual stimulation. Affect is the end result of a perceptual-cognitive process involving the awareness of uncertainty in relation to the possibility of coping with a novel situation. This awareness of uncertainty, occurring while a person is in a state of normal consciousness, manifests itself in the so-called "orienting reflex" (Sokolov, 1960) and is experienced as arousal. The experience of arousal is neutral in quality, i.e., neither pleasant nor unpleasant, or both pleasant and unpleasant. But the "orienting reflex" is likely to elicit a process of self-appraisal, which is what gives rise to an affect, pleasurable or unpleasurable, depending on the outcome of the self-appraisal.[1]

The existence of a neutral state of consciousness such as the experience of arousal was recognized by introspective psychologists long before it became identified by neurophysiologists as a function of the brain stem, of the ascending reticular activating system, also known as the "arousal system." As early as the middle of the last century, Bain (1859) spoke of "neutral excitement" or "feeling as indifference." As he put it:

> We may feel, and yet be neither pleased nor pained. A state of feeling may have considerable intensity, without being either pleasurable or painful, such states are described as

[1] According to Pribram (personal communication), the orienting reflex or reaction has two components: a searching-alerting one, and one of registration-memory (short memory).

> neutral or indifferent: surprise is a familiar instance. There are surprises that delight us, and others that cause suffering; but many surprises do neither [p. 13].

Freud's (1926) concept of anxiety as a signal affect may also be seen as referring to the same phenomenon, the experience of arousal corresponding to an "orienting reflex" and serving as a monitor of behavior, a motive for self-defense. Viewing signal anxiety as the experience of arousal that accompanies the awareness of a novel (in the sense of unsettling, incongruous, or conflictual) situation might allow for a better distinction between signal anxiety as an unconscious, "thoughtlike" (Schafer, 1964) affect, and anxiety as a conscious, painful experience that results from the ego's preconscious appraisal of the self.

Commenting on the James-Lange theory, Kubie (1941) has pointed out that it is possible to experience all the somatic components of an anxiety state without any subjective experience of anxiety. In his view, the secondary and tertiary visceral and somatic changes that characterize an affect are incidental to the experience, "only fellow-travellers of the state of uneasy expectation which is created by the inner struggle between excitatory and inhibitory processes" (p. 271). He defined anxiety as "the ego's reactions to a state of tension which arises in the nervous system. . . when excitatory processes are threatening to break through inhibitory barriers to flood the nervous system" (p. 271). Implicit in Kubie's formulation is the idea that the awareness of such a threat can be automatic, i.e., devoid of psychological meaning. The meaning lies in the phase that precedes the increase in internal tension, in other words, in what creates the internal tension. The sense of danger is conditioned by a previous traumatic event, but what creates tension is a conscious or unconscious cognitive process that involves the self in confrontation with forces that endanger its integrity. It is the recognition of these forces, along with the ego's judgment of its adequacy to cope with them, that comprises this cognitive process, giving rise to a neurophysiological excitatory process that is experienced as anxiety and, conditioned by previous traumatic experiences, forces the ego to take defensive action rather than allowing the excitatory process to break through the inhibitory barriers and flood the system with excitation. Kubie, however, associated the sense of anticipation that characterizes anxiety with the period that follows the arousal of signal anxiety. Once aroused, the process of anticipation may last a long time as a state of uneasy anticipation of "anxiety," which Kubie described as "a state of tense waiting as though on guard. . . an anticipatory warning which occurs in the effort to protect against the outbreak of the basic startle state" (p. 271).

The appraisal of the personal risk involved in the perceptual uncertainty experienced as arousal calls for the awareness of the self as an entity apart from its surrounding world of internal objects—a self-representation that is inherently interactional, capable of action that can effect a change in the object world, a change that may or may not be effective in stabilizing a precarious situation, in removing the uncertainty and sparing the integrity of the self as a subject separate from objects, but is conceived as serving that very purpose. This self-appraisal involves imaginary action—the subject imagining himself as active, rather than being really active—or the loss of the ability to take action.

Time as a Qualifier of Affect

In humans, except during very early infancy or under very special circumstances of stress or emergency, action as image is potential action and as such predicated on the intuitive notion of time—time as future or past, one or the other. This implies the ability to place oneself within a temporal sequence of external events, i.e., the ability to identify oneself (one's needs, wishes, motives, intentions, attitudes, values) as an event, a unit in a sequence of causally related events. Such an awareness can be disturbed in a way that tends to misidentify future events for past and past for future. But, regardless of its objectivity, the ability of a person to imagine the possibility of action establishes the notion of future as a dimension of conscious experience.

Whether such an experience is going to be felt as pleasant or unpleasant depends on the appraisal of the imaginary action in terms of its effectiveness to reestablish control over the perceptual input that creates the internal imbalance (perceptual uncertainty, ambiguity, conflict, challenge), which is "signaled by arousal" (Pribram and Melges, 1969, p. 318).

The effort to control uncertainty without recourse to action involves internal adjustments that operate in two antithetical ways: either by allowing for more stimulation—information input—or by restricting it (Pribram, 1967). One mechanism opens the organism to more information from the environment, to new alternatives or risks in the future, seeking an accommodation with it; the other kind attempts to solve the situation in terms of the past, treats the disturbing stimulation defensively as "noise" (error) and blocks it, focusing instead on intrapsychic events, memories, and convictions oriented toward the past, familiar self-object representations, which ordinarily tend to enhance internal organization.

When the person tries to cope with a situation of arousal by resorting (or intending to resort) to action, affect tends to become nonspecific, its quality reverting to the primary pleasure-unpleasure dimension. So long as one remains inactive, however, and uses neither of the mechanisms described above, one's experience is defined by the here-and-now, a present without future or past; and such an experience carries along a certain affect, which, while basically pleasant or unpleasant, is specific to this particular time orientation. But the more intent a person is to use action, the less specific becomes the prevailing affect, up to the point of being experienced as tension or release, as mere pain or pleasure. Action is dominated by the image of the world of external reality at the expense of the world of internal objects. Self-consciousness or self-awareness and the self in action tend to define two contradictory, mutually exclusive states.

In most instances, at least with humans, an affect is something more than pleasurable or painful—a specific affect like loneliness, longing, grief, boredom, hope, joy. The specificity of human affects is determined by a built-in readiness to experience success or failure, gain or loss, in terms of a self-appraisal that equates the integrity of the self as an internal image or representation with the assessment (cognition) of the ego's capacity to cope with the prevailing organismic disequilibrium in a temporal perspective. As with the pleasure-unpleasure dimension, time is projected outward and felt like an objective attribute of the world. These two dimensions—pleasure-unpleasure and time—combine to determine specific or human affects, which are further qualified by the availability of a familiar, "good" or "bad" object to include in the defense configuration ("neuronal model," "program," "plan") that underlies the affective experience.

Some affects are strictly emotional and some strictly appetitive. But as often, affects transcend this dividing line, involving both a wish and an affective, "emotional" state. When they do so they constitute a different, higher order of affects, an integrated, motivational-affective disposition, which is as exciting as it is fulfilling. Such may be the so-called moral or esthetic affects, Freud's "aim-inhibited" love, conviction (rather than simply drive, wish, or will), self-confidence (rather than simply satiation, fulfillment, or happiness). Conviction and self-confidence may define what Erikson (1956) has conceptualized as "sense of identity" and, in their absence, "identity defusion."

Time and the self, time and consciousness, affects and time are pairs of subjective reality, phenomena that appear together in the course of man's ontogenetic development and define human nature as distinct from that of all other living creatures. Even though conceptualized independently, they cannot be experienced separately, they cannot exist without each other. They can be felt with variable intensity, depending on internal needs and external circumstances, on one's particular state of mind, the one or the other—time, self-consciousness, affect becoming more or less dominant in our awareness, more or less conscious, more or less important in our experience, momentarily or for much longer. They define normality as well as pathology. To put it differently, more conventionally, they are influenced by one's normal or pathological state of being—character structure, states of emergency, drugs, meditation, altered states of consciousness, psychiatric illness.

V

TIME AS A DIMENSION OF AFFECTS

THE EXPERIENCE OF TIME as future, past, or present, time as tense or perspective, is as ubiquitous and characteristic of the human condition as is the experience of duration, time as flow, motion, or change, as a fellow traveler who moves by quickly or slowly, or stops, or from time to time gives us a ride and makes us forget that anyone or anything is moving or changing. Both experiences have in common the notion of change, but when experienced as moving, time is perceived as something or someone apart from the self; in the experience of time as tense, time is perceived as part of the self, as a property of the self rather than of the object world. For when we speak of the future or past or present as something outside of ourselves, we are no longer referring to our experience, to what we feel or live, but to a concept that helps define our existence without being experienced as such. If we try to conceptualize time in terms of its tense, we are likely to arrive at an image of time as space rather than as an object that moves or stops in space.

Anxiety and Depression

As Abraham (1911) observed, "anxiety and depression are related to each other in the same way as are fear and grief. We fear a coming evil; we grieve over one that has occurred" (p. 137). What relates these affects to each other is a temporal frame of reference. Anxiety, like fear, refers to a "coming evil"; depression, like grief, to "one that has occurred." Psychological or experiential time is a qualitative determinant of affects.

After developing his structural theory of personality,[1] Freud (1926) defined anxiety as something that takes place in the ego or, more properly speaking, the self.[2] In perceiving a danger and the possibility of being overwhelmed by adverse internal or external forces, we experience anxiety, which then serves as signal for action, for defense against the danger. Freud postulated that anxiety, as the ego's automatic reaction to danger, is conditioned by ubiquitous infantile situations of helplessness, beginning with the threat of loss of mother, mother's love and that of other love objects and, eventually, the threat of castration. Freud, however, emphasized that anxiety is experienced when the ego perceives a threat and a conditional helplessness, the possibility of being overwhelmed by an event in the future.

As Waelder (1967) spelled it out, "the ego anticipates future events and samples the unpleasure of a future catastrophe in small doses—the anxiety signal—which then, through the pleasure-pain principle sets the avoidance action into motion" (p. 17). Anxiety is brought about by the anticipation of a dangerous event, overwhelming in its implications but still avoidable.

Taking Freud's idea one step further, Bibring (1953) suggested that depression, like anxiety, is an experience triggered by the cognition of an event of catastrophic significance; only in the case of depression this event is perceived as inevitable, as

[1] I like Janet Malcolm's definition of Freud's structural theory as it appeared in the *New Yorker* (November 24, 1980, p. 87): "Conceiving of the mind in terms of three psychic agencies: the ego, the id, and the superego, which stand for reason, passion, and conscience, and whose fate it is to be locked in perpetual conflict."

[2] Speaking of the ego as the seat of anxiety and affects, as Freud (1926) does, is awkward, if not wrong. The ego is also the agency that utilizes affects as signals for action. Thus, it is passive and active at the same time, both subject and function. Such a conceptual ambiguity explains why contemporary psychoanalytic writers have found it necessary to define a special entity within the ego or superceding the ego, namely the concept of the self (Hartmann, 1950; Jacobson, 1964; Kohut, 1971). Concepts such as "ego identity" (Erikson, 1956) or the "realm of experience" (Sandler and Joffe, 1969) serve the same purpose. One should then talk about the "self" or the "realm of experience" (conscious or unconscious) within the ego as being the seat of anxiety and affects in general. As Levin (1969) put it: "Anxiety over a danger in self-object relations is experienced within the self; in the course of its development the ego comes to take this anxiety as a signal to set defenses in motion" (p. 48). Then the experience could be one pertaining to self-esteem and its fluctuations, or to a "level of safety feeling," or "basic feeling state" (Sandler and Joffe, 1969).

unavoidable. More specifically, one experiences depression when one perceives oneself in a condition of irrevocable helplessness—not only threatened, but defeated, doomed. As Zetzel (1960) put it: "While anxiety may be defined as the ego response to a disaster which threatens, depression represents its response to one which has materialized" (p. 480).

Those who imply that anxiety and depression are conditioned by the subject's orientation in time may find the idea too obvious to elaborate.[3] Yet the presence or absence of a time dimension in experience is crucial for the understanding of anxiety and depression as well as other affects—all specific affects beyond the two elementary generic ones of pleasure and unpleasure.

Perceiving Oneself as Adequate or Inadequate

As already pointed out, the concept of time refers to two different realities: one that is defined as mathematical, rational, or "thought" time, and another that stands for subjective, experiential, or "lived" time (Bergson, 1907; Reichenbach, 1951). It is the latter that Bertrand Russell (1929) obviously makes reference to when observing that "The importance of time is. . . rather in relation to our desires than in relation to truth" (p. 21). What determines the "lived" time's position along the continuum of future-present-past is not the cognition of external reality, but rather the way one perceives one's own state of adequacy (what Freud and Bibring identified as the ego's experience of helplessness) with reference to some adverse or challenging reality, within or without oneself, to which one's attention becomes obsessively cathected.

Actually, no matter what the position of the external event or its mental representation on the continuum of time, the feeling of time is always in the present. As St. Augustine (Augustinus, *ca.* 398) put it long ago:

[3] Brenner (1979), whose elementary text of psychoanalysis has served well many beginners in the field, has elaborated on the idea, but failed to acknowledge its origin.

> . . . if there are times past and future, I wish to know where they are. But if I have not yet succeeded in this, I still know that wherever they are, they are not there as future or past, but as present. . . . Wherever they are and whatever they are they exist therefore only as present [p. 257].

The traumatic event the ego perceives may be real or imaginary, external or internal, future or past, but it is invariably felt in one's self as intrasystemic tension in the present. This experience—whether anxiety or depression—is immediate, perceived by the dynamic system of the self—the ego—as a painful event in the present.

As academic psychologists (Schachter and Singer, 1962; Piaget, 1962) and psychoanalysts (Schur, 1969; Ross, 1975) have demonstrated, the role of cognition in the determination of affects is paramount. In order to experience a certain affect as such, one must first perceive a given situation as relevant to one's own well-being, to one's needs or aspirations, and on the basis of past experience assess one's capacity to cope with it. If one's ego assessed itself as incompetent, and the self as vulnerable, the result would be a painful tension within its system—variously described as a decrease in self-esteem (Bibring's [1953] narcissistic core of the ego), as a state of helplessness, or as a breach in the ego's boundaries (Ramzy and Wallerstein, 1958)—but not yet anxiety or depression or anything as specific as that. What determines whether this tension will assume the specific quality of anxiety or depression is the ego's orientation in time, which in turn is determined by the ego's assessment of its own state of adequacy as potential (future) or actual (past). Sullivan (1954) expressed the same idea in his own unmistakably personal way:

> The problems presented in handling anxiety are, it seems to me, vastly clearer if, whatever your initial predilections or previous training, you come to accept my definition of anxiety. As I use the term, anxiety is a sign that one's self-esteem, one's self-regard, is endangered. That is a sign which occurs with strikingly prospective quality—that is, anxiety is often a sign of foreseen lowering of self-esteem. In

> this sense, anxiety is very smooth-working, and usually not very disturbing, for it usually precedes that which would disturb self-regard, indicating by its appearance that a change must be made in the progression of activity—even if only in the progression of thought—in order to insure the maintenance of an appropriate regard for the self. In other words, anxiety is a signal of danger to self-respect, to one's standing in the eyes of the significant persons present, even if they are only ideal figures from childhood; and this signal, other things being equal, leads to a change in the situation [p. 218].

To the extent that the ego can perceive the noxious influence—in essence, the ego's own inadequacy—as something that is still in the future, only potential and therefore avoidable, the experience tends to be that of anxiety rather than depression. It is the hope or awareness of the probability characteristic of the sense of future that gives to the affective experience the quality of anxiety. As Flugel (1955) cogently observed: "Apprehension as to what is to come is the great price that man has to pay for his power of foreseeing the course of future events and adapting his conduct so as to influence them favorably" (p. 192). The following case illustrates the relationship of anxiety as an affect and the subject's particular orientation in time:

> The day before I was going to leave for my summer vacation, a patient of mine complained that she had a headache. "I feel awful," she elaborated. "I have a feeling like something bad is going to happen, a feeling as if something is going to happen in the world—more than a feeling about what is going to happen to me. It seems bigger, more global. I have this feeling of apprehension about what is going to happen, I don't know how to explain it, may be just a larger concern that just—I don't know, it's exactly like I am afraid the whole world is going to blow up, as if I am afraid about what is going to happen to the future . . . I definitely feel it's a feeling of insecurity, of something about the future. It's mostly that I am concerned about your going away."

Conversely, by perceiving itself inadequate in relation to a noxious event in the past, the ego assesses reality as inescapable, unavoidable, irrevocable—finality being the essence of the past. And the ensuing experience is that of depression. The following vignette illustrates the affect of depression as related to the sense of time:

> While helping his father mow the lawn, a young boy caught his hand in the blade of the motor. Even though no fingers were lost and the surgeon who treated the boy ascertained that everything seemed to be in order, the boy's father felt depressed and remained so for several days. The memory of the accident kept returning compulsively to his mind along with the realization that he had anticipated the danger, but too late to prevent it from turning into a disaster. He thought of all the various ways the accident could have been avoided, blaming himself for not instructing the boy better, blaming his wife for having suggested that he let the boy help him and for insisting that they should pick up the cut grass right away, a thought that brought back to mind the fact that he himself had used his hands to pick small masses of grass dangerously close to the blade of the machine in front of the boy, who had apparently imitated him. He remembered thinking that it might rain that afternoon, in which case he would not have been able to do the work and the whole thing would have been avoided. He remembered the day before, riding with the boy on their bikes to watch the trains maneuvering in their tracks, and wished he had done the same that afternoon, instead of mowing the lawn. It was as if he were trying to find an escape hatch, something to take him away from what happened that afternoon, only to be faced with the inevitability of the event. He felt trapped by the memory of the event as if that were the only reality and everything else subordinate to it, unnecessary, inevitable, and irreversible antecedent. He felt lost for being unable to find his way out of it, a stranger in a land where there was only one sense of time: that of the past.
>
> While in this state of depression, he could continue his work, moving and communicating in the present and planning for the future, but nothing in the present or future held any appeal for him. And unless he made a concerted effort to resist the temptation, his mind was at any moment ready to return to the past.

If, however, the ego becomes convinced that the damage is actual or certain, an impending calamity may result in a depressive affect rather than anxiety—as if the damage has already been done, even though, objectively, it may still lie ahead and one knows it. If the awareness of the possibility that the catastrophic event may be avoided is weak or absent, in other words, if one's ego loses the ability to perceive the possibility of escaping defeat, the noxious event is no longer perceived in the future, but as something already in the past, and one feels depressed—as in the following case:

> A middle-aged woman, in psychotherapy for a number of years, had periodic attacks of anxiety characterized by an excessive concern about money. She knew that her financial resources, modest as they were, could sustain her celibate and frugal existence for a long time. Yet she would look into the future and feel anxious, imagining that something might go wrong and leave her penniless. She was advised to get a job, but nothing available was ever to her liking. Then, following a series of adversities, none of which affected her financial situation, she began to complain of increased anxiety and also of depression, becoming so handicapped that hospitalization was recommended. Even though she could easily afford to be a private patient, she insisted on being admitted to a state hospital, claiming she was poor—a notion she justified with the argument that she could not expect to leave the hospital before all her money had been spent. In talking about her financial situation she gave, indeed, the impression of being already destitute. She described herself as hopeless and helpless. Her suffering was obviously actual and urgent, but her affective experience was oriented exclusively toward the past. Even though she was concerned about the future, she could not conceive it as such, compulsively identifying it with the past.

According to Schmale (1958, 1964), the affect of depression consists of two component feelings: helplessness and hopelessness, comprising the "giving up-given up complex." These feelings reflect awareness of the ego's inability to cope with a loss of two different types of autonomy. Helplessness reflects a loss of ego autonomy and the sense of deprivation resulting from a loss of gratification ordinarily provided and expected from a nurturing object. Hopelessness reflects a loss of ego autonomy and the sense of despair resulting from the awareness of one's own inability to provide the desired gratification. Genetically, the sense of helplessness may be traced to the late oral period of childhood development, when the establishment of object constancy enables the infant to identify the maternal object as the source of gratification, as well as frustration; while the sense of hopelessness has its origin in the phallic period, after the self-concept has developed sufficiently to include the superego's influence on the notion of time as future.

> On his first hour back from a business trip, a patient in psychoanalysis talked about a long, perplexed wandering in the streets of Washington, D.C. as he was trying to find his way from the convention hall back to his hotel. He recalled how he became scared at the idea that the city was populated mainly by blacks and that it was getting late, but he did not feel anxious until he found himself lying in bed, alone in his hotel room. "Then I started feeling cold and anxious," the patient recalled. "It was the fear of aggression perhaps, plus the insecurity of not knowing what you were going to do ahead. Something triggered it as I found myself alone in the hotel room. It was something intolerable. I felt isolated, cold and shaking all over. I wished to call my wife, or a girl I had met at the convention, but it was 2:30 in the morning—it would sound stupid. I was feeling anxious and completely isolated, feeling that you could not call, like a baby waking up in the middle of the night and feeling there was nobody there." At this point the patient recalled a scene from childhood, mother taking his two older brothers to school and leaving him alone in the house. "I must have been very young," he speculated, "as I was still in my baby bed—I can still see myself standing up in my baby crib, crying, scared." Returning to his recent experience at the hotel room, "I was feeling very anxious," he elaborated, "fearing of becoming helpless, fearing that you wouldn't be able to manage it. . . feeling terribly anxious, sort of not knowing how to alleviate that anxiety, like sometimes if I'm anxious or depressed I'll have a drink and alleviate it, or I'll go out and have contact with somebody, or work in the yard, or masturbate—like that night, I got up, took a shower, masturbated. It was being like a baby, an infant, feeling cut off of all support. Perhaps it had something to do with the group experience early in the day, when I felt really loved by people, sort of elation, high as a kite. And all of a sudden to find yourself alone—I laid there and rehashed the

> meeting, going over back and forth our arguments. . . I was also approached about becoming the president, arguing back and forth until I took over the mike and told people to shut up and that I would recognize only one at a time and that had quite an effect on the group. That night in the room I was thinking about that and how there was a good possibility I would be president next year—which makes me anxious. I was thinking, here I was like an infant, anxious to death, feeling helpless and hopeless, when a few hours earlier I was acting like a supremely self-confident, perfectly responsible adult."

In general, a loss or separation which leads to feelings of helplessness and hopelessness may be real, but also threatened, expected, or fantasied, symbolic, internal or external, conscious or unconscious. Depression results from the ego's inability to perceive a replacement for a real or fantasied loss.

Boredom and Depersonalization

In discussing the mechanism of depression, Bibring (1953) pointed to a comparable, phenomenologically allied affect: boredom. More than any other affect, boredom is experienced as a disturbance in the sense of time, as an inability to synchronize attention with the activities of the surroundings or, in their absence, with one's own fantasy life. The experience generally involves a sense of impatience with the self and the environment; a sense of frustration, dissatisfaction, and want; a vague need for something, for an interest or outlet. Even though it may not impress the observer as much as the suffering that depression or anxiety usually convey, it is an unpleasant, painful feeling. In metapsychological terms, boredom results from a partial selective repression, wherein the ideational representation of sexual or aggressive drives is repressed (the object or goal of the wish is unconscious), but the drives or wish can still be felt. As Fenichel (1953) put it: "The instinctual tension is present, the instinctual aim is missing" (p. 293). In other words, boredom is the result of the awareness of one's instinctual impulses without the awareness of their meaning.

What one experiences in boredom is a defeat, a sense of inadequacy that has all the attributes of a finite failure, but which is not projected temporally, either into the past or into the future. One's ego is able, for the time being at least, to check the threatening impulses so well as to deny them all meaning, including the meaning of danger and future. At the same time, the ego is not able to use its energy in any context, to pursue any goal, being bogged down by the dangerous forces. The individual feels defeated; his ego boundaries are breached. But again the experience of defeat, occurring without cognitive context, is devoid of all meaning, including that of the past. As his inadequacy—the breach in his ego boundaries—cannot be placed in a time perspective, future or past, the individual experiences neither anxiety nor depression, but a sense of impatience with the present.

Commenting on the subjective time peculiar to the experience of boredom, Fenichel (1953) recognized the possibility of a primary disturbance of the ego regarding the dimension of time, a disturbance which might facilitate the emergence and operation of what he considered as the mechanism of boredom, i.e., the repression of drive-aims and the vain search for substitutes in external stimulation that is only too remotely related to primary instinctual aims. In pathological cases of boredom, such as that experienced by individuals with a borderline personality organization, the picture is much more complicated, reflecting a deficit in internalized object relations and a defective superego.

> A young woman began her analytic hour by expressing the wish to take a trip, perhaps to Nepal. She was not interested in cultural things when traveling, she said, and became bored very quickly in museums. She felt as though she ought to be interested in things in a museum, but she just got very tired. Then she talked about how unreal everything in the world seemed to her. Things fifteen minutes in the past are just shadows of what they were, and things fifteen minutes in the future can barely be comprehended because she cannot imagine that they would be real. She said it was safer that way. At this point, her analyst noted that the patient's voice became "very flat and monotonous."

The suggested affect here is one of boredom, with overtones of depersonalization. Bibring (1953) has pointed out that these two affects are related to depression, and that all three represent states of mental inhibition. He distinguished boredom from depersonalization on the basis of the kind of underlying instinctual drives or derivative wishes—drives and wishes actively blocked from consciousness—libidinal in the case of boredom, aggressive in the case of depersonalization. An ominous kind of barely suppressed aggression is evident in the following case:

> "I am having an unreal feeling," a patient of mine said after lying silent for some time on the couch, "as if I am getting my days mixed up. . . . It is as if something is missing, as if I am waiting for the ax to fall, somehow an unreal quality." Then, after some more silence: "I somehow feel as if I am churned up inside, but not in a way that hurts. I feel kind of fluid, almost it seems it ought to hurt, but it doesn't. It has kind of a timeless quality. I am usually conscious of time as I feel I am always behind; now it has a timeless quality, something unreal. . . . I feel as if I am going to be out of control; I feel as if I am going to hurt. It is a good feeling, it's just the opposite of going out of control, but I keep asking 'what am I overlooking?' It is too good to be true." He suddenly recalled a series of associations he had yesterday as he found himself sitting behind his analyst at a doctors' meeting and feeling very comfortable, in contrast to other times when he usually felt bored or absent-minded, he was feeling comfortable for no significant reason, when he noticed that the analyst's hand looked jaundiced. Before he realized that this was an illusion caused by the artificial light, he had a series of associations, going back to the time when his father got jaundiced and went fast to his death with terminal leukemia; after a brief period of happiness for everybody, everything had gone from right to wrong in the space of one week. And now he began having idle fantasies of some disaster, his family all being killed in a car wreck, or his house, wife, and children wiped out by a tornado. "Maybe I am looking for a disaster . . . I don't know if that is wishful thinking or fear," he muttered.

According to Bibring, what distinguishes boredom and depersonalization from depression is that in the first two "the self-esteem of the person remains outside the field of conflict" (p. 32). The contention that a person who experiences boredom or depersonalization does not suffer a decrease in self-esteem seems unjustifiable, however, when one considers the fact that both boredom and depersonalization are unhappy experiences, even though for some individuals preferable to depression or anxiety. The fact that the quality of experience in boredom and depersonalization—in comparison to depression and anxiety—and the degree of suffering involved in these two sets of affects are different may be better accounted for by the difference in the way the individual experiences himself and his inadequacy with reference to the time continuum. In the preceding examples the subjects, describing their discomfort either as boredom or depersonalization, find it difficult to experience time as either past or future. They experience themselves in a slow-moving, almost immobile present, which they obviously do not enjoy, yet find preferable—"safer"—than either a past or a future orientation, which might bring about depression or anxiety. Significantly, both patients were prone to severe anxiety attacks.

It has been observed that the sense of time becomes disturbed when the person's identity and set of values are vulnerable or become threatened. Disturbed adolescents experience "time diffusion," as Erikson (1959) put it, "a loss of the ego's function of maintaining perspective and expectancy, [a condition in which] every delay appears to be a deceit, every wait an experience of impotence, every hope a danger, every plan a catastrophe, every potential provider a traitor. Therefore, time must be made to stand still. . ." (p. 141).

A person may be future-oriented and still feel bored, finding his occupation monotonous or meaningless, as a present-oriented person would tend to do. Indeed, a future-oriented person may lack imagination and the ability to project his self meaningfully or attractively into the future. He is future-oriented but in a fragmented, disconnected way, the future into which he projects himself having nothing to do with the present that keeps him captive. The present for such a person is extraneous, having no organic, meaningful relation with the future except as a sample of what is to come—a continuous, interminable, unchanging present that leads nowhere but to an end that makes no difference: death of body and soul. Time, when it does not contain the conditions of change—change for the better or for the worse—is experienced as slow, monotonous, boring. Time goes fast for people who perceive their present as inherently related to their future, what they are doing now and what they are experiencing now as leading to an emotionally arousing future—a future that makes a difference, good or bad.

Other Negative Affects

Like anxiety, depression, and boredom, other affects, painful as well as pleasurable, can be placed into a cognitive relation involving the ego's self-assessed adequacy and its position along the continuum of subjective time. For the purpose of demonstration rather than because of any intent to account for the entire spectrum of affects, I shall continue with negative, unpleasurable ones for a while, using as a frame of reference the triad of anxiety-boredom-depression and their corresponding position on the continuum of time. In the process, I shall refine my conceptualization of affective experience so as to distinguish affects from each other when they have reference to the same time orientation.

An affect of central importance for psychoanalytic psychology is guilt. Freud (1930) defined it as "the perception which the ego has of being watched over [harshly by the superego]. . . the assessment of the tension between its own strivings and the demands of the superego" (p. 136). Anna Freud (1936) defined it simply as "superego anxiety," the affect experienced when one is threatened by his conscience. More specifically, Fenichel (1945) recognized a temporal element in guilt when he found it necessary to distinguish between feelings of conscience which judge the past, and guilt feelings which warn about the future. As he pointed out:

> Guilt feeling proper—that is, the feeling "I have done wrong," a painful judgment about some past occurrence which has the character of remorse—must be distinguished from feelings of conscience which do not judge the past but

> the future. "I should do this," or "I should not do that." This part of conscience has a *warning* function and directs future actions of the personality [p. 134].

Indeed, the subjective quality of guilt will differ depending on the ego's position along the dimension of time as it perceives itself to be inadequate in a conflict that involves the superego. Such a perception could place the ego's impairment in the past and therefore render the affect of guilt an experience similar to depression. The fact that guilt and depression are often experienced concomitantly is indicative of a common temporal orientation. On the other hand, the possibility of breaking a prohibition imposed by the superego and thus facing the threat of punishment on the part of the superego would place the impairment, so far as the ego's assessment is concerned, into the future. As Spiegel (1966) observed: "Anxiety deriving from the anticipatory function of the superego necessarily has a quality of futurity attached to it. It may take the ordinary normal form of general concern about the future" (p. 330). Such an experience might be expected to resemble that of anxiety rather than depression. Whenever the superego is involved in the conflict that confronts the ego, anxiety or depression acquire an overtone of guilt.

If the sense of time is indeed a qualifier of experience when one's ego is confronted by a traumatic event, then referring to the affect that results from the superego's punitive reaction for a past performance as guilt, and to the affect that results from the superego's warning about a future performance also as guilt, presents a problem. It would be better if one reserved the term *guilt* for the experience that refers to the future and resembles anxiety, and the term *remorse* for the experience that refers to the past and resembles depression. Guilt has reference to future action, remorse to action of the past. As Emily Dickinson (1890) put it:

> Remorse—is Memory—awake—
> Her Parties all astir—
> A Presence of Departed Acts—
> At window—and at Door (p. 365).

Fenichel (1945) and van der Waals (1948) clearly identified the difference. Some of Melanie Klein's followers, notably Money-Kyrle (1955) and Grinberg (1964), have also referred to two kinds of guilt: depressive as distinct from persecutory, identifying the former with feelings of grief and remorse, and the latter with resentment, fear, and anxiety. Freud (1930) himself wrote:

> When one has a sense of guilt after having committed a misdeed, and because of it, the feeling should more properly be called *remorse.* It relates only to the deed that has been done, and, of course, it presupposes that a *conscience*—the readiness to feel guilty—was already in existence before the deed took place [p. 131].

Freud further suggested that "the term 'remorse' should be reserved for the reaction after an act of aggression has actually been carried out" [p. 137].

Another affect related to anxiety is fear. Fear, of course, implies an ego orientation toward the future. What distinguishes fear from anxiety is that the former occurs in a situation of danger where the threat is distinctly perceived as external, i.e., in outer reality. The ego is able to outline the danger, either because it is exogenous, originating in an external object or event as in a wild beast or a natural disaster, or because it is displaced onto an external object or event, as in phobias. In contrast, anxiety occurs when the dangerous forces are primarily endogenous, intrapsychic, and only vaguely defined, as in the case of a threat from the id or the superego. In the face of danger, it is the ability of the ego to commit itself to a course of action in coping with the danger or, rather, with the perceived effectiveness of the action, which determines whether the prevailing affect will be fear or anxiety. As Epstein (1967) has pointed out, the more ambiguous a situation of danger, the more difficult it is to perceive the possibility of coping with it by engaging oneself in directed activity, and consequently the more prone one is to experience anxiety rather than fear. Otherwise, both fear and anxiety refer to affects that result from the ego's assessment of its adequacy in terms of the future, the ego perceiving a state of helplessness which may or may not prevail—uncertain, avoidable, something in the future.[4]

Arising from the loss of a valued object when the latter remains external and distinctively identifiable, grief is to depression what fear is to anxiety. Like depression, grief results from the ego's assessment of itself as inadequate with reference to something that has already happened, something that belongs to the past.

Among other basic affects, fatigue and anger—like boredom—are experienced when the ego is unable to project itself and its inadequacy into time, future or past. But, in contrast to boredom, the forces that cause the inadequacy are identifiable, either because they are not repressed, or because they are external. Actually, what is identified as the cause of fatigue or of anger may not be the true or original cause of these affects, but rather some substitute object against whom these affects can be displaced and without whom no such affects can be experienced—only undifferentiated tension. As is true for drives, affects tend to have an object, which is identified as the cause or source

[4] According to Heidegger (1927), fear is an affect that is determined by the sense of the past, by the dimension of the past rather than the future, even though the emotional situation may involve danger in that sense, potential and not actual harm, harm in the future. In contrast, the sense of dread or anxiety is with reference to the future, because the self projects itself into the future without concern for its present or past security. But, even though what Heidegger describes as anxiety or dread may be conceivable as an experience, it represents a metaphysical situation, as Heidegger himself would agree, rather than a psychological reality.

The paradoxical thing with Heidegger is that he talked about experience as if it were not a psychological phenomenon, and about moods as if they did not refer to affects, even though he named them as such, depression and elation in the case of a state of mind that is concerned with actual reality, with the everyday world of human affairs and material concerns. Yet, is it any less paradoxical when we psychoanalysts talk about unconscious affects or ascribe adult affects to preverbal babies and to the self in its early developmental stages, before the construction of a reasonably coherent ego apparatus, speaking about the ego as a structure of the mind, or shall I say the self? I don't think it would be any easier to talk about the person rather than the self, except that it would avoid confusion between self and self-image or representation, which are of course two different things—or are they? When Heidegger wrote about the awareness of the self, he seemed to refer to self-image or representation, the way one perceives oneself and others in one's private internal world, which is not necessarily a conscious and certainly not a factual, objective world. As Bertrand Russell once wrote of Heidegger's philosophy, "one cannot help suspecting that language is here running riot" (*New York Times,* May 27, 1976).

of the affect—the object or cause of fear, grief, fatigue, anger. Such an object becomes the focus of the ideational equivalent of the affect and provides the opportunity for intentional behavior. But as this object can be replaced by a more convenient one, it can also be repressed, in which case the affect acquires a different quality, similar but not identical to that experienced when its object is identifiable, i.e., within ego awareness. Thus we come to have experiential pairs of affects, such as fear-anxiety, grief-depression, fatigue-boredom, anger-depersonalization, and so on. Schachtel (1959) identified object-bound affects as "activity-affects," and objectless ones as "embeddedness-affects."

In the case of present-oriented affects, inasmuch as the unattainable instinctual, libidinal, or aggressive aim is repressed, the experience in tinted with the quality of boredom or depersonalization. A predominance of libidinal drive gives rise to fatigue; a predominance of aggressive drive leads to anger. This distinction according to the instinctual source of the conflict corresponds to the one made earlier with regard to boredom and depersonalization. Jacobson (1957) makes a similar distinction between sadness—as a reaction to object loss—and depression: "unlike depression, sadness as such does not involve an aggressive conflict, either with external reality or endopsychically" (p. 87).

Positive Affects

The understanding that painful affects are the experiential equivalents of the ego's temporal perception of itself as inadequate can be applied also to pleasurable affects, which may be seen as determined by the ego's assessment—justifiable or not—of its own adequacy in a temporal framework, its success in maintaining the integrity of the self against internal and external odds, in overcoming or avoiding dangerous situations, in growing stronger and safer, in expanding and conquering, in creating. As the perception of danger results in the experience of anxiety and other similar painful affects, the anticipation of gratification or success results in happy excitement that may be identified as euphoria, zest for life, hope. In Jacobson's (1953) words: "The expectation of future gratification appears to induce pleasurable affect components which we are entitled to call signal affects too, though of the pleasurable variety" (p. 64).

At the other end of the time scale, the perception of oneself as successful, the assessment of one's adequacy and the ego's projection into a satisfactory past result in a positive experience, the height of which may be described as elation. The fact that elation has been identified as the defensive counterpart of depression (Lewin, 1950) would justify the use of the term in describing the experience of a sense of accomplishment in terms of the past. Bibring (1953) contrasted elation to depression by pointing out that it expresses a state of "fulfillment." When, however, in appreciating its success, one's ego focuses its perception onto the present instead of orienting itself into the future or the past, an experience akin to boredom should take place, only it would be pleasurable. Such a pleasurable affect without reference to time, future or past, may be identified, and indeed has been described in the literature (Bull, 1951), as joy. According to Goldstein (1939), brain-damaged individuals, having lost the ability to experience time, cannot feel joy in the adult sense. What they experience instead is an affect that may appear like joy, when in essence it is mere pleasure by relief of tension.

Normally, all adult experience is time-oriented; and there is always reference to past, present, or future in varying proportions, but all interacting. As Binswanger (1958) put it:

> The existential sense of temporality in general forbids us to conceive of "future" only as the empty possibilities of the pre-established, the wished-for, and the hoped-for, as it also forbids us to see in the past only that which was present and is now over. Rather we must understand by the past in the existential sense the has-been, which decides that we not only have been but actually *are* from the viewpoint of "has-been." In this "having-been" are founded the "capabilities" by virtue of which the existence exists [p. 302–303].

This is the classical Hartley–Hamilton–James conception of the "stream of thought."[5] But what I would call specific to the

[5]Gardner Murphy, personal communication.

subject of affects is that the ego's particular orientation in time provides an essential element in the qualifications of a particular affective experience, determining to a large extent whether the experience is going to be that of anxiety, depression, elation, and so on. In such a limited and limiting sense, one may speak of time—and I speak here about psychological, lived, or subjective time—as a dimension of affectivity, without implying that time is not a dimension of mental life as a whole.

VI
TIME IN NEUROTIC DISORDERS

Disturbances of the time sense can be traced to unconscious motivation, the activity of unconscious infantile fantasies or wishes, rigid defenses, and the compulsive, prohibitive or punitive nature of the superego. Ubiquitous as they may be under certain circumstances, the experience of slow or arrested time, or the feeling that time goes too fast, are usually indicative of internal conflicts, of an intrasystemic struggle that pits the ego against the id or against the superego. The struggle necessarily reflects one's ego strength and sense of vulnerability with respect to a danger in the future or to a failure in the past.

Time and Obsessional Phenomena

Freud (1909) observed that obsessive individuals, in their compulsive need to maintain themselves in uncertainty and doubt, are known to display a strong dislike for clocks and watches and to employ all sorts of devices "in order to render these doubt-removing instruments innocuous" (p. 232). Compulsive individuals dread the idea of time, which they equate with the idea of death, their own and that of love objects, wishing them to be dead and feeling guilty about it. Compulsive thoughts and rituals have the defensive purpose of suppressing the awareness of the passage of time and the lurking image of death.

According to Stekel (1927), compulsive patients "[do] not wish to accept time. Time is standing still. . . . All these patients agree that usually they do not feel the passing of time. The day is horribly rich with details of compulsive ideas and acts, with doubts and fantasies, so that it is here and gone, fast as the wave of a wand, although many of these patients do not work at all.

Equally frequent are fictitious ideas of immortality and holiness" (p. 64). Another way that compulsive individuals protect themselves from the anxiety-provoking idea of time and its equivalent, death, is by being late for appointments or in meeting deadlines.

An obsessive concern with time may serve as a defense against the emergence of overwhelming instinctual drives and the danger of ego disintegration. The fear of losing track of time represents symbolically the fear of losing one's object world or one's self as a distinct entity, the fear of being thrown back into the past of early infancy and its state of defenselessness with regard to instinctual drives. As an obsessional patient said (see Dooley, 1941):

> Things have to happen in time and I must be on hand to see what happens. . . death and birth can't happen in the same instance. If there is time then I am alive. If there was an interval without time I wouldn't be there! If time ceased, God knows what would happen, I'd be blotted out. . . . I must never forget myself for a single minute. I watch the clock and keep busy, or else I won't know who I am [pp. 16–17, *passim*].

In discussing the symbolic and defensive uses of time made by obsessive-compulsive patients and, in general, individuals characterized by strong anal traits, Abraham (1923) pointed out that the idea of time is universally associated with money. And, as Fenichel (1945) put it:

> Anal personalities are as disturbed in their attitude toward time as they are in the attitude toward money: in respect to time, too, they may be stingy or prodigal or both alternately; they may be punctual or unpunctual; they may sometimes be accurate to the fraction of a minute, and at other times grossly unreliable [p. 282].

The idea was advanced one step further by Jones (1918), who suggested that obsessional neurotics equate time with feces, and that their preoccupation with time, punctuality, and productivity in terms of deadlines is motivated by anal eroticism. As he put it:

> The concept of time is, because of the sense of value attaching to it, an unconscious equivalent of excretory product, and the reaction just mentioned [the fear of castration] is also shown in regard to it; that is to say, people of this type are particularly sensitive about their time being taken up against their will, and they insist in every way on being master of their own time [p. 420].

Harnik (as cited by Schilder, 1942) went so far as to suggest that time in the unconscious stands for the introjected feared and hated father, who is henceforth identified with fecal material. Likewise, Meerloo (1948) ascertained that anal personalities equate time with father and father's demands, the demands of any authority in general. Not being on time expresses, presumably, aggression against the father, and "killing time" acquires the symbolic value of killing what is experienced as the tyrannical force of authority. Thus an obsessive-compulsive patient with a strongly ambivalent attitude toward her father said: "time is a very relentless master, time is boring, the only thing you can do is to kill the time with masturbation" (Meerloo, 1948, p. 603).

With their punctuality, compulsive people end up by becoming slaves of time, organizing their bodily functions—eating, defecation, coitus, sleeping—more in relation to time than to inner, instinctual needs. Such people end up feeling guilty for spontaneous drives and activities. They are continually aware of the existence of time, which they try to schedule in a meticulous way as a defense against the fear that aggressive wishes against the father whom time symbolizes will become dangerous, kill him, or provoke his revenge.

Lewin (1950) assumed that neurotic as well as normal people come to identify time with father, time becoming a metaphor for the father and the reality principle, the adult expectation to renounce one's infantile pleasures for the sake of adaptation to reality. And to the extent that mother stands for the pleasure principle and timelessness, "time as the father is the intruder into the timeless relation with the mother" (p. 310).

For some neurotics, being late produces autoerotic tension (Harnik, as quoted by Meerloo, 1948). Writers and artists have a tendency to delay their creative activities, being very conscious of the passage of time, which they experience with enjoyable tension and a sense of omnipotence. According to Meerloo (1948) again:

> Unless the patient clings to time and the clock, repressed fear and aggression break through. He fights against free creation because, in full creation, he is unaware of time. He steals time from others in his tediousness; he tries to obtain full control of somebody else's time. For time is also a symbol of life. Giving time to somebody is loving him. Giving one's own time and attention you are giving oneself. People who "never have time" and who hide behind their duties, do not love. Sometimes in analysis, time finds itself symbolized as the quintessence of love, as an orgastic experience—"those two minutes in which I really lived." At that moment, time stops [p. 606].

In more benign cases, the obsession with time may stand for the missed good mother of childhood, as implied in the associations of one of my patients, an obsessional young man who complained that he had constantly the feeling that time was limited, that he could never finish anything, could never have enough time to do all the things he wanted to do; analysis would be over before he had what he wanted, which brought to mind how different things were in childhood, how he could complete a game and go home where mother was waiting with food and loving care—time was limitless then. She used to be after father for coming home late but she did not mind with the children.

Morris (in press) presented the case of a young hysterical woman with strong obsessional features, whose experience in handling time served as "the organizing schema for her life, character pathology and transference neurosis." The fantasy of omnipotence inherent in her ability to control time, her own and that of her analyst, dominated the treatment process, creating a major resistance and at the same time serving as a stabilizer of the therapeutic alliance. The case is worth quoting in some detail as it traces the origin of the patient's preoccupation with time to early object-relations, reenacted during several years of psychoanalysis in terms of time-related concepts and behavior.

> The patient, a single professional woman in her early twenties, complained that she could not maintain a fulfilling relationship for long enough, becoming preoccupied with hypochrondriacal concerns after each loss and the fear of "either dying at an early age before I've begun to really live or becoming a vegetable and having to be taken care of forever." She added, "I feel like I'm stuck in a rut. The same things happening to me, and nothing has changed in ten years."
>
> While in college, she was diffusely anxious most of the time, always behind in her school work and usually late for classes and dates. As she said, "Everybody was upset with me because I was either late, procrastinating, or rushing around. I felt like time was running away with me, that my life was rushing by, but somehow I wasn't living it. At least when I went home on weekends, my mother would give me good advice about how to manage my time, and my father would help me catch up on my homework."
>
> Time became an issue in her analysis right from the beginning. In a display of a strong superego pressure that also implied the neurotic wish to be told what to do in order to rebel, she said: "Analysis takes a lot of time, and I don't want to waste time, so I'm only going to talk about what's important. Where should I start?"
>
> She came regularly and on time for the first two weeks, then later and later, and was not on time again for the next six years. When confronted with her chronic lateness for analysis, she laughed: "So what. I am the patient. You have to wait for me," she said, adding: "I'm not so sick that I need to be here on time, and if you would say more, I wouldn't have to come at all." At other times she remarked, "I would lose my identity if I obeyed you and came on time"; or, "there is always plenty of time left. I'm not missing out"; or, "I can come as long as I want. No one else gets my place until I say so." The fear of being controlled or of losing control was expressed usually in terms of time—whose time it was, who controlled it, who allowed it, and so on. Once, when her analyst was late, and even though she herself had come again late, she reacted with rage, threatening to terminate "if you ever do that to me again. This is my time," she said, "and you're supposed to always be here. If you are going to waste my time by

starting late, I'm not coming. You owe me two minutes, and I'm not going to pay my bill until you make it up to me."

Her obsessive need to control time was somehow associated with an uncanny ability to monitor the flow of time, at least in the analytic situation, without recourse to external measures of time. Both wish and ability, manipulative and disruptive as they were, reflected childhood concerns and experiences with time, especially as they involved her parents. Mother and father were themselves obsessively preoccupied with time and used it explicitly as a means of controlling their children's behavior and each other.

Concerning the vicissitudes of the perception and, in general, experience of time, Stekel (1912) wrote:

> The significance of time for the neurotics is more clearly seen in their unconscious calendar. Their dreams betray the fact that in their minds they are continually juggling with certain numbers, especially the years of the birth of certain persons, the years of their death, the days of certain traumata, etc. Gloom and depression on certain days prove on analysis to be the anniversaries of certain important occurrences [p. 445].

Pollock (1970, 1971a and 1971b) has dealt extensively with anniversary phenomena, which he attributes to an uncompleted or pathological mourning process. In anniversary mourning rituals, the ideas of time and death are hypercathected in the way phobic objects are. As with phobias, this preoccupation serves the defensive purpose of warding off anxiety aroused by a conflict which is tied to the time and content of the original traumatic event.

Depersonalization Experiences

Disturbances of the time sense can be traced to the activity of unconscious omnipotent fantasies. As Scott (1948) has pointed out, the severely neurotic tend to engage in an unconscious manipulation of time, rendering it meaningless by reversing it, speeding it up, slowing it down, making it jump, or stopping it. On the other hand, as Scott (1970, unpublished) has also suggested, the ego seems to take account of the steady progression of the world around it, either as a "going-on-itiveness" or as a "particulate now, now, now, now," which is then translated into the secondary-process language of objective, shared time as timelessness or eternity.

To illustrate his point, Scott presented the case of a neurotic woman whose distorted sense of time he traced to unconscious fantasies of omnipotently controlling the duration of her parents' sexual intercourse, to which she had been witness as a child. Similarly, a patient of mine was able to associate the incidence of brief episodes of depersonalization, from which he had suffered since childhood, to masturbatory wishes and the memory of his parents' bedroom. The connection was suggested to him when such episodes recurred during impulsive visits to his own apartment in the middle of the day on the vague suspicion that his wife, normally away at work, was back home with a lover. On the basis of additional associations, it was possible to make the reconstruction that as a small child, while witnessing the primal scene, the patient had imagined himself in his mother's place, in possession of his father's love and penis. The intense self-awareness that made him feel as if he were "almost pure mind," and the "terrible excitement" that characterized his experience of depersonalization, were hence interpreted as a revival of the infantile experience of union with father, while the feeling that time stood still was explained on the basis of the original omnipotent wish to prolong the experience indefinitely. Significantly, while describing the experience as extremely disturbing, the patient confessed that he valued it as something that proved his uniqueness, his "specialness" and "superiority" as an individual—in other words as something that sustained him when his narcissistic defenses were threatened, as in moments of loneliness and oral frustration that were dominated by fantasies of envy with reference to his wife, analyst, and mother.

According to Schilder (1935), who cited Nunberg and others as being of the same opinion, "Almost every neurosis has in some phases of its development symptoms of depersonalization" (p. 139). The experience is often defined broadly to include a subjective change in the perception of one's own self and the surrounding world. But the perception of one's environment as unusual —for which Meyer-Gross (1935) established the term derealization—may occur alone, without disturbance in the sense of one's personal identity. Derealization is probably identical with the phenomenon of *jamais vu*—usually described as the negative of the experience of *déjà vu*—the feeling of strangeness with something that is well known to the individual and should normally appear familiar.

Depersonalization (or derealization), and *déjà vu* (or *jamais vu*) are phenomena that involve a unique disturbance in affect and time sense. In many cases they blend in what is felt like an uncanny experience. Depersonalization and *déjà vu* (and correspondingly, derealization and *jamais vu*) are probably interchangeable, depersonalization involving more of a sense of unreality with regard to affect, *déjà vu* more with regard to time. In both instances the individual retains an adequate sense of personal identity but feels a strange distance from it, as if his self is disembodied, suspended in space or time or both. In *déjà vu* the environment and the events a person is experiencing strike him as familiar against the fact that they are new. The person has the vivid impression that he has been there previously, when he knows that this is not really so. As Schilder (1942) put it: "The world seems far away. Time does not move, and whatever happens seems to be a repetition of what has happened long ago" (p. 214).

In the case of depersonalization, there is a felt distance between oneself and one's feelings, with little if any sense of false recognition. Frequently reported by neurotics, both experiences may occur in normal people under stress (Leeds, 1944; Richardson and Winokur, 1967).

Freud's (1936) report on his "uncanny" experience on the Acropolis established the defensive nature of the disturbance. In a brief article he wrote on the occasion of Romain Rolland's seventieth birthday, when he himself was going on eighty, Freud recalled how one summer, more than thirty years earlier and rather unexpectedly, he and his younger brother Alexander had found themselves, tourists in Athens, on top of the Acropolis. There, instead of experiencing "delight or admiration," Freud was seized by surprise at seeing that things he had heard about as a young boy in school—the Acropolis and its surrounding landscape of Athens—did really exist. Actually, he felt as if split, one part of himself surprised to find out that something he had been taught about so long ago was indeed real, and another part surprised at his having ever doubted the idea. Analyzing the situation further, Freud understood that coming to Athens made him feel guilty, for it proved him superior to his father, who not only did not know any better but was not able to afford the luxury of a trip to Athens—a disrespectful idea he carried along with him, repressed since childhood. Depersonalization, Freud concluded, was a defense against the awareness of an old, guilt-ridden memory.

Freud certainly went much further than his father had predicted when he, as a little boy of seven or eight, had the audacity to urinate in the parental bedroom. "That boy will never amount to anything," Freud remembered his father saying scornfully on the occasion (Jones, 1953–1957, 1:16)—a humiliating memory that, along with the awareness of his many accomplishments, came back repeatedly in Freud's dreams, and was likely to breed vengeful wishes.

The experience of depersonalization, with its sense of detachment from actual reality, protects the self from an upsurge of aggressive wishes and persecutory fear reactivated by the present surroundings. In Freud's case, oedipal wishes and feelings were revived by the sudden awareness that he had proved more successful than his father.[1]

Schilder (1942) attributed depersonalization to narcissistic, voyeuristic, and sadistic wishes or fantasies. Elaborating on Oberndorf's view that depersonalization involves erotization of thinking and a conflict between feminine and masculine tenden-

[1] Reconstructing the episode on the basis of contemporary documents, Schur (1969) suggested that Freud's self-doubts, as manifested in his experience of depersonalization, had more to do with his ambivalence—still unconscious at the time—about the competence and credibility of his intimate friend, Wilhelm Fliess, rather than his father.

cies, he cited the case of an obsessional neurotic, in which episodes of depersonalization served the purpose of protecting him against the awareness of the fear of death, which he associated with the idea of continuous torture and annihilation. In the patient's unconscious fantasies, eternal destruction and mutilation without time limits reigned as the price he had to pay for escaping death, a partial expiation for murderous fantasies. When analyzed, these fantasies turned out to represent homosexual sadistic impulses toward the father. In the case of another obsessional patient, also cited by Schilder (1942), time was experienced as moving too fast. In this case, the patient's obsession with the idea of death was found to be caused by aggressive tendencies "toward life"—presumably masculine, assertive, ambitious fantasies motivated by the superego rather than by prohibited destructive impulses—which could not be fulfilled.

According to Arlow (1966), depersonalization is the experience of observing oneself in action. Pointing out that depersonalization is often associated with the feeling that time is overly extended, he reiterated Schilder's view that it serves as a defense mechanism in reassuring oneself that the threat from one's own unconscious aggressive fantasies, or from some external danger, may lie in the distant future.

An interesting case of depersonalization that combined a serious disturbance of affects and a phobic attitude toward time has been described by Eissler (1952). The case was characterized by the mechanism of isolation, which amounted to an inability on the part of the patient, a woman in her early twenties, to acknowledge any strong feelings or to experience any events of the past as being related to the present. Even though she talked about affects, they did not seem to have any impact on her. When obviously angry, she usually had a smile on her face; when complaining of feeling depressed, she sounded entirely unconvincing. She was horrified by the concept of time and tried to ignore its passage. She actually felt there was no real progress of time, she had "no feeling of the stream of time." She said: "Time has nothing to do with me." And yet she was scrupulous about keeping appointments, always worrying that she might be late. She experienced her life as a disconnected series of events that made her feel as if she were amnesic.

In trying to account for his patient's pathological mode of time experience, Eissler considered three alternative possibilities: a defense against anxiety over some infantile conflict, an early malformation in the development of the concept of time, or a regression to an early mode of time experience.

Disturbances in Time Perspective

The existence of anxiety is predicated on the awareness of danger and the anticipation of failure in the future. But the greater the anxiety, the less conscious a person is of the future as such; either he defensively denies the future in favor of a more controllable present or past, or he suddenly panics, unable to distinguish a threatening future from a catastrophic present.

> "While on the couch yesterday," said a patient of mine, "I had suddenly a feeling of alarm, the fantasy that my heart had stopped, was slowing down or stopping, a feeling of lightheadedness that made me want to get up and walk up and down. It was like my depersonalization attacks, but centered on the chest this time. And now I had the fantasy that I was going to die, which has never been part of my depersonalization attacks before. If this was an anxiety attack, it was never before as severe." Then, in going over the content of the last hour, he recalled talking about his plans to move his business to another state in the near future and the possibility of terminating his analysis prematurely. And that was when he suddenly felt like dying (or causing his analyst to die).

The feeling that time is slow or not moving or endless and that the world is meaningless, empty or populated by disinterested objects while the self exists alone, loveless, purposeless and unworthy, is likely to be identified as boredom. Boredom, and its extreme form, emptiness, is an affect to which narcissistic personalities are particularly susceptible. The feeling of slow or endless time corresponds to an inability to place oneself in any other time perspective than the actual present. Past or future become like incomprehensible properties, prohibited areas that one cannot trespass. An example of the way the affect of boredom is conditioned by the subject's particular time perspective was given in the previous chapter. The following is an elaboration of the same case:

> A young woman in psychoanalysis described her weekend as relaxed but too long, too slow, and at the same time as if it never happened. She wished she could take a trip, do some traveling, perhaps to the Far East. She was not interested in cultural things when she went traveling, and she confessed with a little bit of embarrassment that she became bored in a museum very quickly. She felt as though she ought to be interested in things in a museum, but she would just get very tired, very bored. Then she thought it might be interesting to visit the International Exposition in Montreal. She usually was not interested in architecture, but was told that the architecture at the Exposition was very exciting. She then talked about how unreal everything in the world seemed to her. Things fifteen minutes in the past were just shadows of what they had been, and things fifteen minutes in the future could barely be comprehended, for she could not imagine them as real. Yet she felt safer that way; there was no pain when things were unreal. Now she began to criticize herself: "Why do I do things like that?" She felt like a factory worker on an assembly line, all her life doing things she did not really want to do. She felt angry and helpless. At this point she broke down and cried. She then went on to say she had let her mother organize her life for her since she was a little girl.

Disturbances of Time as Duration

The experiential length of a given time period seems to depend on the relative urgency of one's impulses or, as Anna Freud (1965) put it, "on the measuring being carried out by way of either id or ego functioning" (p. 60). Instinctual impulses are inherently incapable of delay and waiting. It is the ego that by the mediation of thought introduces postponement of action, which is as characteristic of the ego as urgency of wish fulfillment is of the id.

> How a child will experience a given time period will depend therefore not on actual duration, measured objec-

> tively by the adult, by the calendar or by the clock, but on the subjective inner relations of either id or ego dominance over his functioning. It is these latter factors which will decide whether the intervals set for feeding, the absence of the mother, the duration of nursery attendance, or hospitalization, etc., will seem to the child short or long, tolerable or intolerable, and as a result will prove harmless or harmful in their consequences [pp. 60–61].

And, sooner or later, the development of the superego adds one more controlling factor. As a patient of mine said:

> I don't know if this has to do with psychoanalysis, but my perception of time seems to be all screwed up in the last three months. Time seems to pass too fast. . . . And I'm so often angry, it makes me angry to look at myself, the more I look at myself the more I find I need to work on, and I don't like that. I don't like my vague thinking and some of my passive-aggressive self; the more I am in psychoanalysis the more problems I see in myself. . . . For some reason, I find that my day passes before I realize it is over; and I don't get some things finished that I would like to do. . . . It's almost as if my focus for the day is to get to analysis and. . . I started to say, get it over with, get to psychoanalysis and another day will be gone. And I don't say this in order to please you. . . on another level I feel I want to come to analysis and learn as much as I can.

Time appears to move slowly when instinctual wishes remain unfulfilled or become repressed. Time appears to move rapidly when wishes are converted into or replaced by superego demands and expectations. Let me illustrate the point with one more clinical vignette.

> Mr. A., a man in his middle thirties, was again late, by four or five minutes, for his analytic hour. He apologized. When I invited him to associate his tardiness, he remarked that he was usually late for all appointments—at work, on social occasions, etc. He thought this was because he did not want to be considered a "good guy" and a "compulsive worker." He recalled being a "good student" in high school, but not wanting to be known as the "teacher's pet" and therefore "taking it easy" with assignments. It

> was the same here, he felt; he did not want to appear anxious to do what was expected of him. To be late was part of himself; it was the way people knew him, and he did not wish to change. He hurried to say that he was aware of the implication such a wish had for broader change: he was probably "resisting" giving up what he felt was being himself. He talked about "taking it easy" at first, feeling that he had "plenty of time" and then hurrying up at the end as he realized that he was "getting late." Then he expressed two wishes: to be able to "take it easy" so as to enjoy what he was doing, and for this he needed to "hold time back"; and to be able to do as much as possible within the time available, and for this he needed to "rush with time."
>
> He felt as if time was going too quickly, "slipping by," and he could not catch up; his moments, his days, his years went so fast that soon he would be dead and not able to accomplish what he wanted in life. On the other hand, it seemed to him as if his "internal living" was too fast and the outside world too slow, because it could not offer him what he wanted to have immediately.
>
> "It's like a spinning wheel inside my guts, as if being on a treadmill," he complained. Then, in a more reflective tone, he said: "Not that the world is slow, but my productivity, my effectiveness is slow." He recalled his mother's speed of action—"she talked a lot. . . yakity-yak. . . it upsets me to think of it, and while yakity-yakiting she was doing a thousand things, baking cakes, going through a million books. In my inner turmoil I'm equal to my mother. Even father was more productive, like he would come home from work and tear into a million things. In one evening my dad could do a thousand things more than I can do; my mother would digest half a book. It takes me a long time to get on to tasks, I'm wasting a lot of time. . . will I be able to do as much as Dad did, will I read as much as Mother? She moved quick too, she was always doing things. I have this feeling. . . will I get it all done, or will I screw it up?"

It would seem that this patient experienced the flow of time with regard to himself in two different, apparently contradictory ways: on the one hand, as something moving slowly and himself as running; and on the other, as something going fast and himself as getting late, falling far behind. These alternate views regarding the flow of time corresponded to two conflicting orientations in his personality, both reflecting internal motivating forces. When motivated by his id (the urge to express himself and to enjoy things in life), he experienced time as something slow, unfulfilling, boring. When motivated by his superego (the need to achieve, to be efficient and productive like his parents), he experienced time as something fast, elusive, worrisome.

Time as duration — the experience of time as slow or fast — is a function of the strength of our instinctual drives and a metaphor for the integrity of the self in relation to the world of internal objects. To the extent that instinctual drives are pressing for discharge, creating a dangerous situation for the ego — or, to put it in the framework of object-relations theory, jeopardizing the autonomy and cohesiveness or integrity of the self-image — time becomes of essence, being experienced as something moving or stationary. The speed of the apparent motion is inversely related to the strength and urgency of the instinctual drives or instinctual response system: the stronger the drive the more slowly time seems to pass, and vice versa. The taming influence of the ego and the substitution of instinctual drives by goals originating in the superego tend to create the impression that time goes more quickly or, in extreme cases, so quickly as to make one feel left far behind, even immobile. The impression that time has stopped, on the other hand, seems to be the result of the ego's desperate, last-ditch effort to defend the integrity of the inner self-object world from threatening instinctual, libidinal, or aggressive drives.

VII
TIME IN BORDERLINE DISORDERS

THE BORDERLINE PERSON IS CONCERNED exclusively with the present, with the here-and-now, with the wish to be loved immediately, with the hurt and anger he feels when he cannot have the love he wants right now. He is ready to promise anything, to throw away his own life and those of others, to throw away his future by drinking hard, by taking pills, by shooting himself or others, oblivious of the consequences, aware only of his anger—a feeling that knows no future, compresses all future and all past into the present moment.

The Borderline Syndrome

Borderline disorders comprise a syndrome that refers to five general aspects of human functioning: affective disposition, impulse control, reality testing, personal identity, and interpersonal relations. In this general framework the borderline patient displays a number of specific, pathognomonic characteristics, which in turn make for a diagnostic profile that transcends traditional classificatory labels.

The borderline individual is full of anger. His anger is characterized by oral demands, which has often a paranoid flavor, and is expressed in hateful outbursts against anyone who happens to frustrate or disappoint him. Besides anger—which is what impresses the observer most—the borderline person experiences an array of vaguely definable dysphoric states of mind, affective dispositions, or moods. Boredom, a sense of emptiness or depression—something that lacks depth or despair and has been described as "affectless" depression (Grinker et al., 1968)—and free-floating anxiety are most prevalent. Anxiety may be experienced as tension and discharged in impulsive, seemingly purposeless, unpredictable action.

The borderline individual has a diffuse personal identity, being often sexually promiscuous or perverted, prone to suicidal gestures and other self-defeating behavior, including psychotic episodes, which are characteristically brief, easily reversible, and as a rule associated with some identifiable external stress like abuse of alcohol or drugs. Ordinarily the borderline's reality testing is adequate, but it suffers in the absence of external structure, as can be readily demonstrated by projective techniques like the Rorschach and the TAT.

> An obsessional, basically borderline patient was taking the Rorschach test. After giving twenty fragmented and more or less idiosyncratic, highly ambivalent responses to the first card, she turned to the examiner and said: "I feel like breaking your stopwatch." Then she laughed, as if to reassure him, or perhaps herself, that she was not going to misbehave, implying that she needed the boundaries of time in order to maintain her reality testing, threatened by the unstructured nature of the test.[1]

Above all, the borderline is noted for the intense but shallow nature of his interpersonal relations. Even though he is eager or secretly wishing for human contact,[2] his poor sense of identity and defensive, basically narcissistic disposition make it difficult for him, if at all possible, to maintain friendships or to experience intimacy and love, other than fleeting infatuations.

Bizarre phobias or obsessions and a hypochrondriacal preoccupation are often present, as a rule more than one at the same time. Such symptoms blend with the affective disposition of the borderline person to form various character types, more or less disordered behaviorally as well as subjectively, beginning with their sense of selfhood or ego identity. The dynamics of the dis-

[1]Siebolt Frieswyk, personal communication.

[2]One possible exception is the schizoid patient, at least as the new *Diagnostic and Statistical Manual* of the American Psychiatric Association (DSM-III) describes him. It might be noted that Fairbairn (1941), who adopted the term from Bleuler (1916), saw in the schizoid personality a relatively benign borderline condition.

turbance include a set of regressive, narcissistic defenses, mainly splitting, denial, primitive idealization, and projective identification, which can be detected by psychological testing and are readily activated in psychotherapy or any intense emotional confrontation (Kernberg, 1975). The idea that borderline disorders have a common genetic or developmental origin is more difficult to demonstrate; but it has heuristic value, being helpful when it comes to treatment and the understanding of the patient-therapist relationship.

An Array of Present-Oriented Affects

A heightened affective pressure, based on a large quantity of primary narcissism and unbound aggression, makes borderline people keenly aware of the passage of time but unable to experience affects other than anger and its equivalents, i.e., disgust, envy, and omnipotent elation; or boredom and its equivalents, i.e., feelings of emptiness, alienation, helplessness, and hopelessness. These are present-oriented affects, bound by the here-and-now, affects devoid of time perspective, future or past.

> A young college man came to his psychotherapy hour looking disgruntled, absent-minded. With some encouragement on his therapist's part, he started talking about a play-reading group he had just attended. He said he was angry because he had an "angry" role there. He was supposed to be angry and he got involved in this role. It was T. S. Elliot's *Cocktail Party,* and he was reading the part of Edward, the husband whose wife is telling him that he is not alive, that he is passive, and so on. The patient realized that was the way he himself felt; and the realization made him feel very angry to hear someone say the right thing about the way he himself felt. He was not feeling "alive." I asked him to elaborate, and he said he felt he was not a "true person," he was "not alive," he did not feel like living in reality; it was as if time passed and left him behind, or as if he were waiting for something to happen, as if he were sleeping while an alarm clock was ticking the time away. The alarm was going to go off at any moment, yet it didn't, which made him feel as if the alarm clock was a "time-bomb."

The anger of borderline patients is presumably the result of the awareness that some previously experienced or anticipated gratification and the gratifying object, as well as the activity, are unavailable (Schmale, 1964). Anger turns into apprehension if the unpleasure persists or increases as the individual is moving away from or reducing his activity. Developmentally, negative affects such as those borderline individuals experience *par excellence*—anger, fear, helplessness, and boredom—become differentiated from anxiety when the sensorimotor activities of the child fail to provide gratification. In particular, helplessness as an affect is likely to occur whenever active search for, or active avoidance of, external objects does not bring pleasure or does not relieve unpleasure. The affect of helplessness, undisguised or experienced as boredom, anger, or depression, is likely to recur later in life whenever there is a loss or threat of a loss of an object that is expected to supply immediate gratification and there is no way to replace it or overcome its loss. Most often the loss is largely imaginary, representing no actual absence of a gratifying object, but the internal unavailability of an early object like mother and the object loss sustained during the separation-individuation period that follows the so-called symbiotic phase of an infant's development (Mahler et al., 1975). The following letter of a young borderline woman addressed to her hospital therapist illustrates the regressive nature of such feelings:

> I hate myself after realizing in therapy today that I want you to love me, to make me feel important by treating me like you would treat a friend who is not a patient of yours. I know I shouldn't feel this way. When you are nice to me I want to have you stay near me and never leave, and when you say goodbye to me everyday it tears me up inside. That's why I want you to be angry with me so it doesn't hurt so much inside me. No, I don't want you to be angry with me—I just want you to understand why I sometimes want to turn up my anger on myself—it's because I want to punish myself for not adhering to what is best for me. . . . I want you always to be by me and yet at the same time I know you cannot take care of me as you would your own child or

wife. This is such a hard reality for me to face, but I face it every day and by facing it I feel bitter and rejected.

A sense of unfulfillment and the wish for narcissistic supplies to make up for an early maternal deprivation made this patient feel that time did not move fast enough. And unless she was with her therapist, she felt bored.

The affect of boredom is the experiential equivalent of prohibited wishes and fantasies deprived of their ideational component, repressed sexual wishes and fantasies contaminated by pregenital, predominantly oral, aggression. The experience of boredom is characterized by a perceptual ambiguity, a vague awareness that the self is out of balance, trapped as it were in a field of conflictual forces that allow no option for action, real or imaginary, to restore one's sense of equilibrium and self-esteem. There is a sense of inadequacy that, characteristically, is devoid of time perspective, future or past, a sense of being locked in the here-and-now, in an endless, meaningless, objectless present. People who feel chronically bored manage to wall themselves off from interpersonal relations, often behind obsessional, vaguely sexualized thoughts and compulsive rituals. When acute, as in psychotherapy, boredom signifies inhibited rage against an object that is experienced as frustrating or disappointing—the therapist and the internal object representation revived in the transference.

Boredom is prevalent in the lives of narcissistic and borderline individuals, who are unable to invest themselves in relationships and events that do not offer immediate gratification. Narcissistic patients, in particular, find no gratification in the process of giving, and they like you to the extent that and so long as you are willing to give. Their need for love is one-sided and insatiable, demanding but without faith, desperate in its urgency but suspicious with regard to its reciprocity, to the possibility of being fulfilled, or of being commensurate to the great expectations that borderline and especially narcissistic patients have of a love object. Their unrealistic expectations, reflecting the internal image or an ideal object that has failed to become integrated with the opposite side of the parental image, render the love object impotent, robbing the relationship of human meaning or future. Such patients are virtually in love with love, and from its very beginning the relationship is doomed to failure. Soon, they break off the relationship in disappointment and often contempt, leaving the other one heartbroken, humiliated, and puzzled. As Stendhal (1831), life-long student of human narcissism, put it, "The only way they know how to touch a heart is by bruising it" (p. 27, my translation).

Whether boredom happens to be the experiential equivalent of narcissistic greediness or a defense against the fear that one's infantile rage might destroy one's internal objects, it can be of diagnostic as well as prognostic significance. The borderline patient, whether narcissistic or otherwise, is incapable of investing himself in an object or activity, unless it offers immediate libidinal gratification. He carries with him a structural defect, which is based on the primitive defenses of denial and splitting.

> A sixteen-year-old girl fired a shotgun from the window of an upper floor apartment, killing and wounding several adults and children on the grounds of an elementary school below. She calmly told a newspaper reporter by telephone that she was "doing it for fun." He quoted her as saying: "I don't like Mondays. This livens up the day. I just started shooting for the fun of it." The reporter said the girl's voice carried no apparent remorse or sorrow. "She was obviously a kid," he wrote. "She was calm. She was not laughing." When he asked, "Do you realize that you almost shot two or three people?" she replied, "Is that all? I thought I shot twelve" [*New York Times,* January 30, 1979].

Disgust, another common affect in borderline conditions, is phenomenologically a variant of anger that may pass for boredom: feeling disgusted with something or someone, quite often with oneself. Disgust is a derivative of infantile anal drives and reflects vaguely a superego influence. Envy, or its impersonal variant, greed, is the oral equivalent of disgust. It lies very close to the surface of narcissistic anger, forming in most instances its motivational basis. Anger motivated by envy or greed has a quality of hatred, indicating an excess of oral aggression and a relation to the early anxiety-ridden "paranoid-schizoid position" that Melanie Klein (1946) has described as preceding the guilt-generating "depressive position," associated with the capacity to experience time as future and past, respectively. Grinberg (1977), in fact, has suggested that these two early mental states may account for two types of borderline patients, a schizoid type whose affective life is dominated by hated and persecutory fear, and a less disturbed type, whose affective life is dominated by depression. Envy, greed, and persecutory hatred are motivated by a strong wish to control, and they constitute the affective disposition of projective identification, one of the most primitive and maladaptive defenses that, along with splitting, characterize borderline and severe narcissistic conditions.

The ego suffers from the awareness of strong oral, libidinal, and aggressive impulses, which are denied mental representation and therefore are deprived of time perspective in terms of either future or past. The following vignette may illustrate the point:

> After reporting a series of recent successes, professional as well as social, a young man in psychoanalysis for two years commented: "In spite of all this recognition, I left the hour yesterday feeling depressed, as if analysis had lost its meaning. And I had the same feeling today while in the lounge waiting." Then the patient recalled paying his analytic bill and the analyst not acknowledging it, which inexplicably made him feel as if he had lost something, and then a vague feeling of depression descending upon him, not serious enough to have fantasies of death, but making him want to resume his old pattern of staying up late, overeating, not doing his work, watching TV instead, and being lazy. He experienced a feeling of disengagement and apathy, a suspension of all sense of responsibility. He felt as if time had stopped. And as the hour came to an end, he felt that he wanted to stop coming to analysis, that the analyst was bad because he had not interpreted a dream fragment he had brought in that day, all of which left him with a feeling of despair that things would go on and on without end, without change.

Disappointed for not being rewarded when he brought in the money for his analytic bill, this patient experienced a loosening of his libidinal ties to the analyst, perceiving the analytic situation as meaningless and endless. As he could not bear to admit his rage at the analyst, he experienced a loss of interest in him and in his own progress, an apathy and a feeling of hopelessness that became projected onto the object world as an arrest of time—the sense of present without future or past.

Identity Diffusion and Depersonalization

The complaint of emptiness and its equivalent expression of hollowness, deadness, nothingness, or inner void, the feeling that "something is missing" or "bad" inside, are offered with tiresome persistence by borderline patients. Such expressions of affective disturbance have been traditionally interpreted as deriving from the unconscious memory of deficient mothering experienced as "empty breast," "mother is not there" (Greenson, 1953). They are often associated with "a decided disbelief in the possibility that time may bring change, and yet also of a violent fear that it might" (Erikson, 1956, p. 82).

Borderline patients suffer from a defective sense of identity, a void in their self-image, a basic deficiency in their self-esteem. In general, the dispersal and fragmentation of affects that characterize the experience of borderlines, in particular schizoid and narcissistic individuals, give rise to a feeling of emptiness, which becomes extreme in schizophrenia—schizophrenic patients often complaining that they have no feelings at all. Acute or precipitous feelings of emptiness may lead the borderline individual to suicide (Singer, 1977). Defensive maneuvers against feelings of emptiness include self-mutilation, sexual promiscuity, compulsive overeating, the abuse of drugs and alcohol, aggressively delinquent behavior and frantic activity in general. According to Isaacs (1948), feelings of emptiness represent a defensive denial of the "bad" mother image.

In contrast to psychotic patients who often complain that they cannot feel anything, borderline patients claim that they have strong feelings which they cannot describe, cannot recognize for what they are, or cannot identify as their own. All they can say when they feel that way is that they are troubled, that their feelings are hurting them, that it is like a hunger for feelings, for love, for attention, for hate. It is a sense of emptiness, but not an emptiness of feelings, as the following excerpt from the notes of a patient with multiple addictions exemplifies:

> I have very strong and disturbed feelings, I have so much to say, and yet I don't know what. Perhaps by writing down any thoughts that come to my mind now will help to get my real feelings out in the open. It's surprising to me that the unhappy can have such a morbid hunger for anything that will increase their distress. This hunger I seem to have had so very intensely and I am ashamed to have to admit this, especially when I tell myself that I want to be happy and yet I do nothing to give myself happiness except for momentary physical, tangible pleasures such as alcohol, or a sudden devouring of food which leaves me feeling miserable afterwards.

The experience of depersonalization occurs with characteristic frequency in borderline patients (Tucker et al., 1973). It is felt as something uncanny and is often associated with a transitory disturbance of the sense of time. The stream of time becomes fragmented and nearly still. This is not the experience of timelessness but, rather, the experience of time taking hold of one's consciousness, permeating one's existence as something discontinuous, extremely slow or immobile within one's self, a sense equivalent to the inner emptiness that borderline persons complain about, which focuses on the loss of the notion of inner space.

As experienced by borderline patients, depersonalization has a quality of omnipotence and elation, in contrast to the equivalent experience of neurotic and psychotic patients, which is fraught with anxiety and is generally unwelcome. A patient with rather pronounced borderline features, in psychoanalysis for some time, said:

> I don't know what it means, but I can picture myself floating into space in suspended animation . . . like I am depersonalized. I don't having any feelings . . . like there is no one there to fuse with, I am all alone. . . . I don't know if that fantasy is a wish ful-

> fillment or not. I don't like being here; I am in some state of regression. . . . It's not unpleasant, but it's not pleasant either, it's just nothing. It's like in my fantasy I see the blackness of space and the stars. It must be what Buddha calls nirvana, a feeling of oneness with the universe.

Anxiety Tolerance and the Future

All affects exist in the present, are felt in the here-and-now. But a number of them are determined by some time perspective other than the present; they are future- or past-oriented. Anxiety and hope are felt in the present but are future-oriented. In order to feel anxious, one must be able to have a mental representation of one's self in a future situation, to assess oneself as adequate or inadequate in terms of a dangerous or challenging situation that is present only hypothetically as a future possibility, as a potential danger or challenge. In order to feel elated or depressed, one must be able to have the mental representation of one's self in a past situation, to assess one's adequacy in terms of a success or failure that cannot be changed, that is "located" in the past.

A "depletion of confident expectation," according to Mahler (1966, p. 167), and a concomitant decrease in self-esteem during the separation-individuation period of childhood development predispose to certain emotional reactions in later life, in particular depressive and anxiety spells, angry outbursts, and boredom. These are typical reactions of borderline individuals. A "depletion of confident expectation" during infancy may more specifically explain the borderline patient's propensity for free-floating anxiety, a state of mind that, even though conceptually related to the future as time perspective, seeks to deny it; to seek the present or the past, instead; to manipulate the time sense for the sake of replacing the very unpleasant affect of anxiety with something that the borderline patient seems to find less unpleasant—boredom or depression, affects corresponding to the sense of time as present or past, respectively. Anxious as he may be, the borderline individual has little tolerance for anxiety and little use for the future as an existential perspective.

On the basis of data derived from the analysis of a patient clearly identifiable as borderline, Oberndorf (1941) concluded that time is an experience inherent in the capacity for reality testing and purpose. His patient, whose progress in treatment was characterized by "compulsive" acting out and "blind" outbursts of "confused rage" alternating with episodes of "profound depression" and "physical inertia," experienced an inability to think consecutively and to associate time with purpose. He made a concerted effort to extend time or to make more time in order to accomplish more, particularly after feeling that he had wasted his time or had failed to do his best. He tried to do too much in too short a time and evaded one task after another. In the effort to control the passage of time, his concept of time would become entirely disrupted and then he would feel "hopeless, uncertain, helpless, unreal or even dead" (p. 151). His loss of time sense, accompanied by feelings of unreality and numbness, seemed to be caused by a strong belief that life had no purpose and therefore was not worth living. He felt obsessed by the sense of time much as he had submitted to his mother's purposes in his earlier life. Time, like mother, had become an autocratic master dominating his life.

If the ability to perceive oneself as adequate or inadequate within a temporal perspective becomes lost, one tends to feel suspended in a pleasant or unpleasant state of mind, which is akin to the putative experience of an infant during the post-symbiotic phase of development, before the establishment of integrated self- and object images, when the self or the object are identified as all "good" or all "bad" and felt as such. Elation, a feeling of omnipotent self-sufficiency, is the positive side of the experience; dejection, a feeling of helpless anger, boredom, emptiness, loneliness, or alienation, is the negative side; and the two sides alternate in rapid, seemingly unpredictable succession.

> A young man, in psychotherapy for several years, described feelings of happiness in direct succession to feelings of depres-

> sion, restlessness, or anger—all in the same hour. At the same time he complained of feeling bored—in fact he did appear completely bored. When I pointed out the discrepancy, he explained that he was talking about feelings he was not feeling right now, but some time ago. He felt bored and he did not know what to talk about to make the therapist feel interested, not to feel bored as he himself felt.

It is very easy, indeed, for a borderline person to shift from one affect to another, from expressing or presumably experiencing one particular affect now and a different one, often its complete opposite, a few minutes later. The readiness of the borderline person to shift from love to anger to hate, from anxiety to boredom to depression, from feeling omnipotently self-sufficient to feeling utterly helpless and alone, is based on a compulsive need for the cognitive restructuring of a given situation or another person. Defensively determined, such a need reflects a severe loss of self-esteem and, on a genetic level, the ambivalent, poorly integrated nature of the borderline person's internal object world. The defense implied is of course splitting, a primitive defense that antedates the establishment of secondary-process thinking and the development of the concept of time. Splitting is a mental mechanism that defies reality testing in one of its most crucial determinants: the ability to time one's experience, to define it according to a temporal perspective. The use of splitting allows one to feel as if one's whole existence is located in the here-and-now, in a way that completely ignores or negates the feelings of the past or the prospect of feeling differently in the future.

> I am living in the present most of the time [another patient said], I am not even fully there. The past is gone, as if it never existed, I can't savor it, I can't do anything about it —somehow it doesn't carry over. I cannot anticipate the future from the past. I am committed to everything but not to a single thing. I don't know what I am committed to— sometimes I say it's me, a real narcissistic character. . . . I don't know about it. I don't feel about myself very well. Here I say I hate the way I am, but at other times I like what I am. I stick by what I am and feel, stubbornly.

> Oftentimes I can't see myself having a single purpose or goal, whatever I do are just tactical shifts. . . . I can see so many things I could be doing, but there is no time for them. Time is all around me, ever-present, but so elusive.

In general, the more one is preoccupied with the future, the more one tends to experience anxiety; and the other way around, the more one feels anxious, the more one worries about the future. As an obsessive-compulsive patient with prominent narcissistic features put it:

> It's something inside me that doesn't let me rest. I am always looking into the future. I say to myself, if only this happened, this thing would disappear. But I know it doesn't. I constantly feel apprehensive, I am worried about being late, about being criticized, about one thing or another. I cannot relax. I am worried about small things; apprehension is always there, I am trying to place it somewhere outside in order to explain it. Like blaming events that are happening on various things and looking into the future for happiness as if to reassure myself that things will be better — I'm worried about the god-damned future, whether something will occur or not occur. What gets me more worried about the future than anything else is seeing myself getting behind in my work and being criticized.

The anxiety of borderline individuals, triggered off by the projection of a vulnerable self-image into the future, triggers in turn a defensive reaction against the perception of the future, as in the following case:

> A twenty-five-year-old man, in psychotherapy for several years, complained about feeling anxious and pleaded with his psychotherapist to relieve his anxiety by giving him drugs. As the therapist tried to explore the reasons that made the patient anxious — such as the feeling that his employer was going to make a homosexual advance or that his parents would get tired of paying for his treatment — the patient became angry and shouted at the therapist not to mention anything that implied what might happen to him, anything that had to do with the future. Then he

apologized, and in a calm, rather colorless way, he said: "I prefer to be depressed; I can't stand feeling anxious. When I feel anxious I have to do something, otherwise I will kill myself or I will die from a heart attack. I prefer to be depressed. I have developed the ability to get depressed—I tell myself 'nobody cares for you, what's the point of trying?'" Looking and sounding generally depressed, the patient continued: "All my efforts have been in vain. I have been in therapy for almost a lifetime, I have had all kinds of jobs. I have tried and failed. I have lost all hope, I have lost all faith in myself, and you, and the ability of anyone to help me." He went on describing his situation in the past perfect tense, as something lost, as a concrete and irrevocable defeat.

Remorse, like depression, an affect with reference to the past, is often preferred to guilt, which, like anxiety, has reference to the future. As Nacht and Racamier (1960) have pointed out, "Remorse appeases the sense of guilt" (p. 485).

Present-orientedness can be induced in anyone by special conditions in the environment, as for example by the circumstances of a battlefield or of a concentration camp (Bettelheim, 1943). Such present-orientedness, being defensive or regressive in nature, is easily reversible. On the other hand, the present-orientedness of a person who grew up under the malevolent conditions of a broken home, whether in a midtown ghetto or in an affluent suburban neighborhood, is more likely to be grounded on a defective, borderline personality structure and therefore more pervasive and resistant to change.

There is a negative relation between the degree of experienced anxiety and the length of prospective time that an individual can mentally tolerate. A borderline patient, recently discharged after a lengthy hospitalization, wrote to a friend as follows:

> I am alone, happy and ever forward in motion and plans, but trying to keep in mind that I am presently in the present. And I am not looking into tomorrow as some other time than the present when it arrives.

The normal individual should be able to act in the present according to what he anticipates to happen in a relatively distant future. But unless the anticipation of the future can give rise to present affects, he is bound to behave maladaptively, to feel and behave like an individual with a borderline personality disorder. The ability to empathize with oneself in the future is probably a function of secondary narcissism or the ability to empathize with another person, an ability that is predicated on the development of object constancy and, in particular, on consistent, unambivalent, well integrated object relations, without which the sense of time is bound to be defective, both as feeling (duration) and as image (perspective), especially in its future tense.

VIII
TIME IN PSYCHOTIC DISORDERS

UNDER NORMAL CIRCUMSTANCES time remains in the background as a preconscious sense, sustaining rather than interfering with the flow of experience; in pathological conditions and situations of emergency, however, time becomes a conscious, dominant sense. This exaggerated awareness of time corresponds to a heightened affective pressure, and is paradoxically experienced as an affectless state. Generated by unconscious fantasies and defenses, distortions in the sense of time and concomitant variations in affect reflect the integrity of one's internal object relations and ego organization.

When Time Stands Still

Failure, defeat, depression, a preoccupation with the past – they all go together and make one feel as if time has stopped. The experience reflects a defensive stance as much as a significant object loss. It is the equivalent of the belief that one has died, that one has no feelings, even though one can still move and think and communicate intelligently. It is like killing one's dangerous, destructive, deadly instinctual impulses and punitive superego to save one's sanity, the semblance of one's self.

The following case illustrates how a subjective loss of affects is experienced as a slowing down or arrest of the flow of time:

> Julia, a twenty-nine-year-old mental patient, complained she had lost the ability to have feelings and, along with that, the sense of time. She referred to her actual experience as "the time world," for everything during the period of her illness, which she dated since a year earlier, when "something inside me snapped and I lost all feelings," centered allegedly on the idea of time. While sitting in her room, she would constantly turn and look at

> the clock to find out if time had moved. She felt that time had stopped and, all the same, she should live forever. Yet, she claimed she had no feelings, that whatever she experienced was not in terms of feelings, but of ideas. She complained of boredom, but insisted this was not a feeling—just the experience of emptiness. Aware of the inconsistency of her claims, she explained: "Everybody says you have emotions because your voice has inflections—but I can't laugh. If I smile, it is a forced smile: it is physically impossible to laugh. But I can cry, even though without feelings. It is always about something in the past, before I lost my feelings, that I cry about."

Indeed, to most people who knew her, staff members and fellow patients alike, Julia sounded "phony." They believed she was dissimulating or denying feelings that were only too obvious to others in her restlessness, in her tears, in her laughter, in the pressure of her speech, in her demands, in her seductive gestures and aggressive outbursts. Julia's history was replete with memories and episodes suggesting the presence of strong sexual and aggressive drives. She was promiscuous, and on several occasions she had made serious suicidal attempts. She herself wrote:

> . . . I am a "lifer" as a criminal; in fact I am a criminal and nothing can change me; things have gone too far. I live in a separate world from everybody else and no one is as lonely or ever has been, except I don't feel a thing since I have no feelings. Nothing moves me, nothing can touch me.

Impressed with her own condition, which she perceived as a unique phenomenon beyond her choice or comprehension, she was eager to describe it in long letters to friends and relatives. She wrote, referring to herself in the third person: "Julia lives in a strange time world."

> Every day is a thousand years, yet the days behind me all collapse into nothing, like a pack of cards. Each day is so long that a normal human being can't imagine it. Each moment is the same way—long. Nothing within this time world has any meaning for me, which is why the time is so long. . . . Morning, afternoon, evening and night are all just the same to me. There are no

natural divisions (the normal person senses natural divisions), but I am grateful for night because when I sleep, time moves without my awareness of it. I know that I was normal one year ago, yet I can't remember what it felt like to be normal. I just know that I was. . . .

Boredom is the key to my behavior. Nothing interests me whatsoever. . . . I feel no energy, no surge of hope—not even depression, in other words, I don't feel.

I have two pasts, the past before I lost my feelings, which my mind struggles to recapture, and the undifferentiated past of my "time world," which does not interest me. The two pasts are separated by the barrier known as "loss of feelings." The past, or before loss of feelings, is steadily fading in my memory, and since my "new past" is meaningless I don't remember it; so, soon, I will have no past whatsoever, and no future; I will exist only in the endless present. . . ."

Depersonalization Experiences

What in her psychotic grandiosity my patient believed to be a unique phenomenon, the French psychiatrist Janet (1928) has described in some detail as *psychasthenia* or *maladie du doute.* He emphasized that patients suffering from such a condition are not able to experience affects. They do not love or hate. Objects appear to them indifferent, or unreal. Pointing out that "these patients have lost the sense of reality" (p. 50), he illustrated this peculiar condition with the case of a young woman who, in the middle of many complaints, said:

> The objects that surround me are unreal. . . . It is very funny, the days do not last longer. . . . I realize that it is evening, I realize that the clock shows such and such an hour, and every time that surprises me because nothing has moved since morning. How's that! Is it time to go to bed? But there is no interval from the moment we get up. Eight hours have gone by, you say, but nothing has changed from the moment before. We are eight hours before, there has been no time inbetween [p. 50].

Janet asserted that he saw many patients of this kind. All of them experienced a sense of emptiness and were obsessed with the idea of time, no longer able to experience the flow of time, feeling that time as duration had disappeared for them, which gave them a sense of unreality, the sense that things were unreal because they could not endure.

Janet's *psychasthenia* seems comparable, if not identical, to what has been described in the literature as psychotic depersonalization, a profound disturbance of affect accompanied by bizarre ideation, hypochondriacal preoccupation, or delusions. The condition may come and go suddenly, but as a rule lasts much longer than similar episodes encountered in neurotic or borderline disorders. There is a strong resemblance to religious conversion, with which it shares a sense of the supernatural or uncanny, except that in most—but not all—cases it is described as intensely unpleasant. Ciompi (1961) has described a remarkable case of psychotic "pseudoecstasy," which lasted for a long time and was experienced as happy:

> At the age of thirty-six, the patient [a married woman of average intelligence with no personal history of mental illness] returned home from church, after a sudden mystical experience, with the firm conviction that the priest S. had looked at her with special significance and would step into her life as lover and become her "fate." Afterwards she stuck firmly to this delusion and awaited confidently and happily "the hour of fulfillment." For two years her daily life was scarcely affected, with the exception of a continuous deterioration of her relationship with her husband. Then, suddenly, acute hallucinations and spiritual experiences appeared. The patient began to hear voices, namely those of the priest S., which disclosed to her precise information about her future happy life. Also, she all at once "recovered her memory," so that she could clearly remember that this same delusional future had been foretold to her twenty years before by a palm reader. She subsequently developed a condition of ecstatic excitement, constantly heard voices, and was finally admitted to the psychiatric clinic in a severely confused psychotic state.
>
> During the initial days [of hospitalization] the patient

lived in a state of ecstasy, carried away by tension that was unmistakably erotic in expression and manner. She spoke constantly of her delusion of the future, which was delineated and confirmed to her without interruption by innumerable auditory, visual, and sensory hallucinations. What was remarkable about these psychotic experiences was the apparent simultaneous appearance and mingling of several temporal levels of experience. The patient experienced her hallucinations, on the one hand, as intensely in the present, and on the other, she repeatedly experienced things that should occur only in the future. On top of all this, she insisted that without exception she had already lived out all her present experiences in the past:

"I am made to repeat everything. I was already here once. Everything in the room was exactly the same. The same people were here and behaved exactly the same as now."

"When I woke up the first morning, I was surprised that I already knew everything. I had already experienced all that once before."

"You already spoke to me once before like this, and I also said the same things as now."

All these hallucinations, anticipating the future, were now also embedded in an absolutely overriding feeling of omniscience. In *visual* hallucinations, for example, the patient saw scenes out of her future life and, at the same time, insisted that she had already experienced all that:

"Pictures were made for me on the wall, like a film, yet in real flesh and blood. I saw my future life with the priest. We lived together. I also saw how we had intimate relations."

"One time we were at a beautiful lake, very happy. Suddenly our child fell into the water. With a lot of effort we were able to rescue him. I see my future experiences."

"All these pictures I had seen once before. I simply had to relive everything."

She reported intense *sensory* hallucinations, in which events in her future were repeated as she simultaneously experienced them as having already been lived once:

"At night I was together with the priest for the last time two hours before my death. I had to sleep with him. I didn't see him, but I felt how he took me in his arms and caressed me. We also had sexual intercourse together. I was very happy."

"The priest made experiments with me. He tried out on me giving birth. I felt myself pregnant and then I had three sharp labor pains. . . . I was made to know that I was living out my future life. . . . But I knew the entire time that I had already once before lived through all this the same as now."

"I had the impression that it took place in the future and in the past. Now I know that it was the priest that mixed the experiences all together."

The *voices* eventually spoke constantly about her future life, which would be foretold in more and more detail. In summary, it was foretold that she would live together with the priest six years, and after a two-year interruption, during which their common child would die, another six months. At that time they would both have to die from a sudden malignant illness.

Such experiences . . . occurred frequently during the first week of hospitalization. However, after instituting suppressive treatment with Largactil [Chlorpromazine] and Phasein, to the patient's disappointment, the hallucinations suddenly stopped. At the same time, her behavior also became more obviously organized, even though for weeks it remained excessively happy and elated. For about four weeks, in every interview, the constant feeling of "having already experienced everything" could be observed, later to reappear for only brief moments, and then completely disappearing. On the other hand, the delusion of a future life with the priest S. that had been present before the acute phase remained unchanged. Despite all this the patient was able to return for a time to her household activities.

After five months, she had to be admitted a second time for a new exacerbation of her psychosis. She suffered again from frequently auditory hallucinations, but this time the state of pseudoecstatic elation was completely missing: the patient was instead grumpy and laconic. No disturbances in the experience of time comparable to those of the first acute psychotic phase were observable [pp. 100–102].[1]

[1] Translated by Dean Collins, M.D.

In most cases of psychotic depersonalization the patient complains bitterly about the experience, while others may not be able to notice any difference in either his behavior or his appearance. The patient commonly complains that he can no longer feel anything, that he has lost the ability to experience feelings. Characteristically, he believes that he is sinful, physically decaying, or dead. When free from such delusional preoccupation, he is likely to complain that time has stopped.

In a series of 26 cases involving the experience of depersonalization, Meyer-Gross (1935) found that 12 were psychotically depressed, six were schizophrenic, and the rest had doubtful conditions that might be diagnosed as borderline. One third of these patients had a prepsychotic personality characterized by hypomanic features. All of them were below the age of 39 and, in their preponderance, women. Their complaints conveyed a sense of estrangement, expressed in terms of a change in the sense of self and in the ability to experience affects, which often involved an altered sense of time. Patients complaining about a change in their selfhood said one or another of the following:

> "I am dead."—"I am a stranger."—"I am different from everyone else."—"I am somewhere else, not here."—"I have changed into someone else."—"I am not a bit like myself, it does not seem to be me."—"A part of myself is not there."—"My mind is apart from my body."—"Something has gone from me."—"My mind seems miles away."—"I am not at all the same, I don't know whether something has come or something has gone."—"I have stopped being."—"I am not the same person like I was before, there are two different persons" [p. 106].

Patients who complained primarily about a lack of feelings described their experience as follows:

> "My emotions are gone, nothing affects me."—"All the feeling is gone, no remorse, no passion, I've only the feeling of being alive, that my heart beats."—"My feeling is dead, I have no interest for my husband and the baby."—"Nothing impresses me."—"I have lost all my love for the children, they seem miles away."—"I am unable to have any emo-

> tions, everything is detached from me."—"The trees in the garden seem unreal, I get no feeling, no thrill, no joy."—"I cannot see the beauty of flowers—the feeling that does not get past my eyes."—"Myself is the same as before, but feelings seem gone away altogether. My body is not changed, not unreal. Lying down I tried to force the feeling back to me. Feeling of remorse, passion, everything has gone, I have only the feeling of being alive" [p. 107].

And those whose primary complaint was that their time sense had changed said:

> "I can't remember my former life, I feel that I have always been in this Hospital."—"It seems years ago that I used to live."—"Streets I knew seem as if I had never been there before. Everything I have done seems such a long time ago." —"When the evening time comes and I think back over the day, it seems years away."—"Time seems like rubber."—"Time goes very slowly, not as it used to be."—"I can't explain, all is timeless, unchanging, hopeless."—"I had no feeling of time at all—time was nothing to me."—"Time just passes, I don't see any future."—"Time seems like a dream."—"I've lost my sense of time" [p. 109].

Meyer-Gross regarded depersonalization as a nonspecific, "preformed functional response of the brain" (p. 118) comparable to epileptic seizures and catatonic states, triggered under a variety of internal or external circumstances and disappearing when the mental disturbance of the patient intensifies, as when depression deepens, or delusion and hallucinations take over. He implicated obsessional character traits. Others (Anderson, 1938; Sedman, 1970) have stressed the association of depersonalization with symptoms of depression.

Variations in the Flow of Time

Borrowing heavily from the work of phenomenologists like Binswanger, Minkowski, and Straus, Schilder (1936) systematically explored the psychopathology of time with reference to psychotic disorders. He pointed out that schizophrenic patients characteristically complain that time "stands still" or that time is "like a piece of ice—silent, as if frozen," that "time does not move" (p. 534). Autobiographical accounts of such patients, beginning with Judge Schreber's (1903) celebrated report that Freud used in attempting to elucidate the dynamics of paranoid schizophrenia, reflect the same experience. Indeed, there is general agreement in the literature that the experience of time as "standing still" occurs in schizophrenia more often than in any other pathological condition (Bouchard, 1926; Fischer as quoted by Israeli, 1936; Scott, 1948; Dubois, 1955; Jaspers, 1959; Ehrentheil and Jenney, 1960).

In contrast, patients suffering from psychotic depression feel that time moves along forever while they themselves stand still. As one of Schilder's (1942) patients said: "I have been here for years. My mother has to live for 2000 years in torment and torture" (p. 216).

Schilder (1936) also quotes one of his depressed patients as complaining that time was going too fast:

> I had the feeling that time was flying. I was here one minute and gone the next. I feel like I am talking to you one minute and talking to someone else the next. I have the feeling that time is flying away. I figure how short time is. We die soon. The pictures in front of me never stay; they are always moving. I cannot concentrate on my work. I feel depressed. The thought of flying depresses me. I know we die in 10 to 50 years but the idea depresses me [p. 532].

A patient of mine, a middle-aged man feeling depressed after a prolonged hypomanic episode, said:

> "Time moves relentlessly, it feels like blows. And yet the spaces in between are so horribly slow. . . . Each day I wake up wishing that I did not have to do anything, to go any place. But time moves on relentlessly. Everything feels like an uphill course, steep and horribly long."

According to Straus (1966), however, depressed patients complain that time no longer moves. Several other workers in the field (Kraines, 1957; Rabin, 1957, Mezey and Cohen, 1961; Dilling and Rabin, 1967; Lehmann, 1967) have reported similar observations, namely that depression correlates with a slowing down of the flow of time. Such a dramatic discrepancy regarding the nature of the time sense with people diagnosed as depressed may be simply an artifact, reflecting an ambiguity of reference in reporting one's sense of time as duration. Like a figure-ground phenomenon, the flow of time may be viewed either as something external to oneself, with the self viewed as ground and the object world as the moving figure, or as part of oneself, as internal movement that contrasts with the outside world, which serves as ground to the self now being viewed as the moving figure.

Whether the movement is attributed to the self or to the object world depends on one's ability to experience affects. Severely ill patients vary in their ability to experience affects and, specifically, the affect of depression. Schizophrenics, in contrast to neurotically or psychotically depressed patients, tend to experience a loss of affects. The denial or suppression of the experiential aspects of their affective disorder results in their referring to time as something "out there," an object in space or space itself that is empty, immobile, or unchanging. Patients who are able to experience depression, however, tend to identify time with themselves, referring to themselves rather than to time as not moving. When they say that time *passes* slowly (rather than *moves* slowly), they indicate that they experience the self as slow or immobile with reference to a world of activities or events which they do not consider their own, activities or events which they find hard to control or identify with. Yet the fact that a person is able to experience affect at all would indicate that he still maintains some ties with the world of internal objects—the notion of time representing the vague awareness of such ties. The tentative nature of these ties tends to create the illusion of time as flying, as moving away from the self while the self remains alone and immobile.

Complaints of depressed patients, inasmuch as they are aware of their depression, convey an overwhelming sense of pastness, a sense of inner immobility that makes the perception of the future as a realistic possibility inconceivable. The self is experienced as immobile while the world around seems to be fleeing, running away, becoming past. The future is overwhelmed by the past, being no longer the future of open choice, but a denial of future as such. As one severely depressed patient put it: "Is there any future? Previously I had a future, but now it shrinks more and more. The past is so obtrusive, it throws itself over me and draws me back" (Schilder, 1936, p. 534). Or, as another depressed patient said: "The future to me is remote. I feel hopeless. Before, I could look for the future, but I can't now. There is something that won't let me" (Straus, 1966, p. 291). Or, "Yesterday seems years ago. I cannot picture any future" (Meyer-Gross, 1935, p. 112).

The Breakdown of the Time Sense

Often, psychotic patients convey the impression that their sense of time is entirely chaotic; moving in an erratic way, now too fast, then slowing down to a halt; becoming interrupted, confused, or entirely absent; without future or past, present-oriented or without awareness of distinctions between past, present, or future. This is what Loewald (1972) has described as the "experience of fragmentation," when the continuum of time —the interrelatedness of past, future, and present—breaks down, and the self is experienced as an empty shell in a world of "bits and pieces, none of which has any meaning" (p. 406). "I had a hard time learning the clock," a young schizophrenic woman told her therapist during a period of acute disorganization. "Is time always going forward or backward? Sometimes it seems it reaches a plateau and stays there. I put my clock in the closet not to see it."

Temporal disorganization is most pronounced in the acute, initial phase of psychosis, in particular schizophrenia, when anxiety is at a peak level. During this acute phase, Melges and Fongerousse (1966) have found that paranoid patients feel as if they are "going faster than the world"; that they live "from minute to minute"; that their attention is "focused primarily on the present" (p. 135).

Temporal organization is a function of the ability to

differentiate between self and object world, inner and outer reality. In an experiment with acutely disturbed psychiatric patients, Melges and Freeman (1977) measured impairment in temporal organization by means of self-assessed statements[2] and found that it correlated highly with Schneider's first-rank symptoms of schizophrenia, i.e., auditory hallucinations, experiences of influence, thought diffusion, delusional perception, and somatic passivity experiences. The authors concluded that temporal disorganization reflects a state of inner-outer confusion, whereby one is unable to distinguish experiences originating within the self from experiences induced by external stimuli.

As Pollock (1971a) has suggested, temporal disorganization may indicate a regressive deficit. Under stressful conditions one's conceptualization of time may regress to an early developmental stage, before the establishment of the ability to estimate duration with any degree of accuracy, to detect succession and temporal association, or to note seriation. Such problems have been described consistently and in the absence of mental illness by individuals who were subjected to prolonged isolation, confined in prisons or caves with limited cues regarding the passage of time. But regressive as any temporal disorganization certainly is, one may frequently detect an element of defense in it.[3]

Psychodynamics of Time Sense in Psychosis

According to Janet (1928), when one examines the psychology of grief and depression, one invariably finds in it a fear of action. On analyzing this fear, one is likely to encounter an

[2] Examples of self-assessed statements used by Melges and Freeman (1977) to measure impairment in temporal organization: "my mind seems to be racing," or "my mind seems to be going slowly" (expressing changes in rate-duration); "my past and future seem to have collapsed to the present," or "my past, present and future seem all muddled up and mixed together" (expressing difficulty in tracking time); and "my mind switches between speeding up and slowing down" (expressing desynchronization).

[3] That the experience of time is not universal may come as a surprise to those of us who were brought up in literate societies. But anthropological data indicate that the concept of time as a continuous process, the ability to project oneself into the future or to perceive oneself as the same when recalling a personal experience in the more or less remote past, should not be taken for granted (Masler, 1973).

anticipatory reaction to failure. The depressed individual goes perpetually through a "reaction of failure." Confronted with the idea of future action, such an individual is likely to say that he would rather not bother with it, as the future can but bring disasters. The notion of failure is furthermore attributed to the external world, exactly like time, as change. Failure comes to a depressed person as something from the outside, as an undesirable event that one is unable to prevent or to turn to one's advantage. Psychoanalytic exploration may uncover murderous fantasies against love objects—fantasies that are handled by denial, externalization, or regressive inaction involving the perception of the self as immobilized in the midst of a world that changes so rapidly as to be rendered absurd.

Arieti (1955) has suggested that in schizophrenia the anticipation of future and in general the conceptualization of time as perspective becomes suppressed, resulting in a "restriction of the psychotemporal field" (p. 247), for the purpose of warding off anxiety. And according to Schilder (1936), the depressive's experience of time as flying and of himself as immobilized is motivated by aggressive impulses and the fear that moving or acting would lead him to commit murder; while the fear of death and the impression that life is going away too fast derives from the wish to reassure himself that his aggressiveness will not be allowed to find satisfaction.

In discussing the case of a young woman whose problems centered on an abnormal handling of time, Orgel (1965) asserted:

> Time and its perception were manipulated by this patient in an effort to ward off the fear of death, conceived of either as traumatic separation from her mother or as mutual destructive incorporation of herself and her mother [p. 102].

Which brings me back to the patient whose account of affect loss and distorted sense of time I presented at the beginning of this chapter. She often expressed the belief that she was a "criminal," that she had hurt her mother and was destined to cause the destruction of her family and the world. While in treatment and following persistent demands for the exclusive attention of her therapist, she became catatonic and attempted suicide. Related to her belief that she had hurt her mother was the fantasy that she had tried to steal father's love away from mother, and a vivid childhood memory of her mother confessing that she had once planned to kill her and then commit suicide. Another early memory, which this patient often recalled while complaining that time and the clock were moving ever so slowly, underscores the overdetermined nature of the experience. As a child of nine or ten she used to leave a note for her father to wake her up in the morning for school, always with some anxiety that he might forget. But each morning her father came—she felt him approaching slowly and then rubbing her back as she lay in bed filled with sexual excitement.

Julia's complex cognitive-affective state, diagnosed as schizophrenia and experienced as a loss of affectivity and a concomitant arrest of the flow of time, served the purpose of defense against the upsurge of primitive drives and drive derivatives emanating from a poorly integrated world of internal object relations.

Unless there is no wish to gratify, time will be experienced as moving—moving faster as the primary, instinctual drives and derivative wishes become weaker and the superego gets the upper hand; moving slowly as one's wishes gain in strength and urgency. If all wishes, including superego demands and instinctual drives, are overcome, time as an experience ceases to exist: one then speaks of timelessness, as mystics and others who claim to have gone through the experience describe it.

One should be careful, however, not to confuse the experience of timelessness with that of time that does not move—the feeling that time does not exist from the feeling that time has stopped. The latter—the feeling that time no longer moves—is the extreme of slow time, when consciousness becomes dominated by the sense of time. Then time becomes one with the wish—time is the wish and the wish is time. As such, time becomes an affect of its own, disposing of all other affects. Indeed, when a person becomes preoccupied with the idea of time, he is no longer able to experience affects in a meaningful way. He is likely to report that he has lost the ability to experience affects at all—to feel, to love, to be joyful or sad or even anxious. Actually, the experience in such a case sounds very much like anxiety; but it is conceptualized as time—the sense that time has stopped. To illustrate the point, I shall quote from another of Julia's letters:

> . . . my soul died last spring—by soul I mean the vital emotions, self, survival, sex, a sense of time. I am completely disoriented in time and space [not in a psychiatric sense, for she was able to identify time and place readily without mistake]. My day is a million years long, in other words I am doomed to live forever—time is subjective and if your soul is dead it stands still. Do you remember Conrad's *Heart of Darkness*? . . . I am eternal, but not in any pleasant sense. Wait until my parents found [sic] out that they have given birth to a "time-space" monster, and you once loved a "time-space" monster. To be cut forever from the world of human emotions and understanding, never to enjoy books, decisions, a sense of time. . . .

Time experienced as not moving is time in its fullest experiential sense. It coincides with the consciousness of the self apart from the world—the self experienced as split from the object world, alone. Time experienced as moving away too fast is only a variation—probably somewhat less pathological—of time experienced as immobile.

When the sense of time comes to dominate the subject's attention, the experienced affect defines a pathological mental state, a stressful situation, or both. The less one is conscious of time, whether as duration or perspective, the more likely one is to be free of psychopathology. Psychotic and borderline conditions are characterized by boredom and a sense of depersonalization, the one or the other or both, affects dominated by the sense of time—time as present. Psychotic depression and morbid anxiety are dominated by a sense of time that is specific to these conditions—the sense of an irrevocable past and a dreadful future, respectively.

IX
TIME UNDER THE INFLUENCE OF DRUGS

MOST IF NOT ALL DRUGS, in sufficient amount, can lead to altered states of consciousness and concomitant distortions of the sense of time. Some drugs, notably LSD, mescaline, marijuana, and the amphetamines, create a heightening of self-awareness and are therefore known as "psychedelic" (mind-expanding) or "psychotomimetic" (simulating psychosis). Others bring about a clouding of consciousness and are classified accordingly as "deliriant psychotomimetics," such as atropine; "inebriants," such as alcohol; "sedatives," such as the barbiturates; "anesthetics," such as nitrous oxide and ether; and "narcotics," such as opium and its derivatives (Balis, 1974).

Time and LSD

LSD-25 (d-lysergic acid diethylamide, a synthetic alkaloid derivative of ergot) induces a compelling sense of immediacy that greatly reduces the importance of past or future (Freedman, 1968; Melges et al., 1971). It is not surprising that this emphasis on the present should create a sense of timelessness (Savage, 1955).

As a rule, LSD slows down the subjective flow of time (Aronson et al., 1959) and eventually, with a peak experience at two to five hours, time seems to stop, or there is a "loss of the sense of time" (Barr et al., 1972, pp. 36–37). At various stages of the experience, subjects typically report that "time is standing still, racing backwards or forwards, or dragging interminably" (Savage et al., 1962, p. 426); or as if "days, months, or centuries have passed" during the few hours of the drug's impact (Klee, 1963, p. 464). The following passage from a long verbatim account of a medical student's experience as an LSD subject gives a graphic illustration of the changes in time perception induced by the drug:

> As I paced, I happened to notice at one point the clock resting on the mantelpiece. It said, I clearly remember, 12:25. Then I lapsed into a train of thought whose various labyrinths seemed to lead me in thousands of directions for thousands of hours. And then I glanced at the clock again. This time it read 12:28, which led me to exclaim "My God, so much has happened in the last three minutes.". . . My sense of time was, of course, strikingly distorted. This may well have had something to do with the sudden unusual speed of my thoughts [Ebin, 1961, p. 360].

Most subjects report or implicitly describe what amounts to an experience of depersonalization or "double consciousness," the feeling that "the body had changed into that of a different person or the self at a different age . . . feelings of unreality, loss of time sense, and loss of reality contact" (Barr et al., 1972, p. 31). An exceptionally vivid case of time distortion and depersonalization that amounted to a typical "out-of-body" experience was reported by Hoffman (1959), the Swiss chemist who first synthesized LSD in the hope of producing an analeptic substance like coramine, a chemically related heart stimulant. Being naïve about the potency of the drug, he took a dose ten times as high as necessary to feel its effects:

> After 40 minutes I noted the following symptoms in my laboratory journal: slight giddiness, restlessness, difficulty in concentration, visual disturbances, laughing . . . I requested my laboratory technician to accompany me home. Since it was war time and no car was available, we went by bicycle. This journey is about 4 miles and I had the feeling of not getting ahead, whereas my escort stated that we were rolling along at a good speed. I lost all count of time. . . .- Space and time became more and more disorganized and I was overcome by a fear that I was going out of my mind.

> The worst part of it being that I was clearly aware of my condition. My power of observation was unimpaired. . . . Occasionally [after arriving home] I felt as if I were out of my body. I thought I had died. My ego seemed suspended somewhere in space, from where I saw my dead body lying on the sofa [p. 245].

A commonly reported finding in experiments with LSD is an intensification of the subjects' emotions, whether positive, negative, or ambivalent. Statements such as "I feel like I am angry—I feel very angry—but I know that I have no reason to be, yet I'm getting angrier by the minute"; or, "I feel like something funny has happened—everything seems funny, but I don't know why" (Katz et al., 1968, p. 11), are characteristic of the experience. Familiar, commonplace, or trivial objects in the environment acquire a novel or "portentous" quality, as in a *jamais vu* experience (Freedman, 1968).

In general, LSD-induced emotions occur without any apparent cognitive stimulation from the environment. Such a finding would seem to contradict the proposition, argued on introspective grounds (Russell, 1927) and supported by a number of laboratory studies (Maranon, 1924; Cantril and Hunt, 1932; Schachter and Singer, 1962), according to which experience needs an appropriate cognitive context in order to become emotional, i.e., to be identified with some specific affect like anger, fear, elation, and so on. But the situation is not unlike what happens when one experiences a strong emotion while dreaming, in which case the stimulation accounting for the cognitive counterpart of the experience is provided by the intrapsychic reality of the subject, the content of the unconscious as it defines and is defined by one's personality structure.

Time and Mescaline

Huxley's (1954) experience with mescaline, so far as the drug's effect on the sense of time is concerned, sounds like that with LSD: "My actual experience has been. . . of an indefinite duration or alternatively of a perpetual present made up of one continually changing apocalypse." Huxley experienced an "indifference" to the idea of time. He thought "plenty of it," but to tell how much was "entirely irrelevant" (p. 21). Summarizing the effects of the drug on himself and others who have taken it under supervision, the popular novelist and iconoclastic thinker pointed out:

> Though the intellect remains unimpaired and though perception is enormously improved, the will suffers a profound change for the worse. The mescaline taker sees no reason for doing anything in particular and finds most of the causes for which, at ordinary times, he was prepared to act and suffer, profoundly uninteresting. He can't be bothered with them, for the good reason that he has better things to think about [p. 25].

A psychoanalyst might be tempted to interpret Huxley's understanding of the effects of the drug on one's willpower as indicating that it improves certain faculties of the ego, especially sensory perception, at the expense of one's superego and motivation in general.

Mescaline is the main active principle of mescal beans or buttons, used by American Indians for their transcendental, hallucinatory properties. The claim that mescaline induces a state of transcendence or agreeable depersonalization has been attested by a number of investigators (Klüver, 1928; Guttman and MacClay, 1936). The following is a graphic account of the experience:

> . . . you have the feeling of swimming in a boundless stream of time, somewhere, somehow . . . you have to pull yourself up repeatedly and make a real effort to appreciate the time situation actively so as to escape from the chaotic flight of time, if only for a moment; only for a moment, however, since as soon as you relax, boundless time returns [Jaspers, 1959, p. 84].

Comparing the effects of LSD and mescaline on schizophrenic patients, Hoch and his co-workers (1952) found that both drugs produced a similar distortion in time sense, generally "a slowing down of the subjective recognition of the flow of time" (p. 582). In some subjects, mescaline induced memories of early life events.

Christopher Mayhew, a member of the British parliament who volunteered to participate in a series of laboratory experiments with mescaline, described his experience in an article entitled "An Excursion Out of Time" as follows (Ebin, 1961):

> What happened to me between 12:30 and 4 o'clock on Friday, December 2, 1955? After brooding about it for several months, I still think my first, astonishing conviction was right—that on many occasions that afternoon I existed *outside time.*
>
> I don't mean this metaphorically, but literally, I mean that the essential part of (the part that thinks to itself "this is me") had an existence, quite conscious of itself, enjoying itself, reflecting on its strange experience, in a timeless order of reality outside the world as we know it [p. 294].

As Mayhew elaborates further on, there was a paradoxical dismemberment of time, a temporal reversal of his thoughts and visual experiences while under the influence of mescaline:

> . . . at about 1:30 all interest in these visual phenomena was abruptly swept aside when I found that *time* was behaving even more strangely than colour. Though perfectly rational and wide-awake. . . . I was not experiencing events in the normal sequence of time. I was experiencing the events of 3:30 before the events of 3:00; the events of 2:00 after the events of 2:45, and so on. Several events I experienced with an equal degree of reality more than once. . . . All I am saying is that I experienced them, not in the familiar sequence of clock time, but in a different, apparently capricious sequence which was outside my control [p. 295].

Mayhew tries to explain at this point that he was in a state of depersonalization, and goes on to describe another typical effect of mescaline on the experience of time, a sense of blissful, ecstatic timelessness:

> The other was even more unexpected and strange. At irregular intervals—perhaps twice every five minutes at the peak

> of the experiment—I would become unaware of my surroundings, and enjoy an existence conscious of myself, in a state of breathless wonderment and complete bliss, for a period of time which—for me—simply did not end at all. It did not last for minutes or hours but apparently for years. During this period I would be aware of a pervasive bright pure light, like a kind of invisible sunlight snow.
>
> For several days afterwards, I remembered the afternoon of December 2 not as so many hours spent in my drawing-room interrupted by these strange "excursions," but as countless years of complete bliss interrupted by short spells in the drawing-room [pp. 295–296].

Time and Hashish or Marijuana

Laboratory studies show that, like LSD and mescaline, hashish and marijuana (extracts of the hemp plant, *Cannabis Sativa*) induce an increased concentration on the present and a reduced ability to project oneself into the future or, to somewhat lesser extent, into the past. The past and the future seem unrelated to the present. This focus on the here-and-now seems to become enhanced by an extraordinary vividness of perception. Thus a subject at the peak of the marijuana effect, approximately one and a half hours after ingestion, said:

> I don't understand it. I don't—there seems to be no transition, no changes in time. It's just *it*. I'm just here. . . right this moment, and I can't—I know I'm moving along, and yet—I'm not. I have no sense to time, I'm just floating along [Melges et al., 1971, p. 566].

Subjects of another experimental study reported that marijuana made them feel that "things seemed to take longer" (Weil, 1969, p. 25). Along with an increase in present-orientedness, there was a discontinuity of thinking and temporal disintegration or impairment of immediate memory. As one of these subjects commented:

> I think I realize why they took our watches. There was a sense of the past disappearing as happens when you are

> driving too long without sleeping. With a start you wake up to realize you were asleep for an instant; you discover yourself driving along the road. It was the same tonight with eating a sandwich. I'd look down to discover I'd just swallowed a bite, but I hadn't noticed that at the time [p. 25].

Likewise, Bromberg (1934), experimenting on himself, reported "little absences of memory." Such an interference with the specious present may account for the feeling that time slows down, breaks down, or stops. As a young man, hospitalized for acute, self-induced marijuana intoxication, recalled: "Time seemed awful long. It felt like the hour would never end. While I walked it seemed that I was going at a terrific rate—almost running" (p. 314). Yet, according to the French poet Baudelaire (1860), the experience is one of timelessness:

> When the following morning you see the daylight installed in your bedroom, your first sensation is profound astonishment. Time had completely disappeared. Only a moment ago it was night, now it is daytime. "Was I asleep, or was not I asleep? Has my intoxication lasted all night, and the notion of time having been suppressed, the whole night has hardly had for me the value of a second? Or rather, was I shrouded in the veils of a sleep full of visions?" It is impossible to know (p. 340, my translation).

Baudelaire's *Les Paradis Artificiels,* ostensibly an account of the effects of hashish, may reflect the poet's experience with opium instead, a drug with which he was much more familiar, especially at the time he wrote about his drug experiences (Ebin, 1961). On the other hand, hashish is ordinarily much more potent than marijuana.

Five years earlier than Baudelaire, the American author and translator of Goethe's *Faust,* Bayard Taylor, wrote about his own experience with hashish: "The fullness of my rapture expanded the sense of time; and though the whole vision was probably no more than five minutes in passing through my mind, years seemed to have elapsed while shot under the dazzling myriads of rainbow arches (Ebin, 1961, p. 47).

Baudelaire's contemporary, Gautier, also writing about the influence of hashish, made reference to time in the context of a hallucinatory experience:

> Upon coming to, I noticed the room filled with figures dressed in black, who went up to one another with a sad air and shook hands with melancholy cordiality, as if afflicted by a common grief. They said: "Time is dead . . . and we are going to the funeral". . . .
>
> "Eternity is threadbare; it needs to be brought to a finish," rejoined another.
>
> "Great God!" I cried, struck by a sudden thought. "If time is no more, when will it be eleven o'clock?". . .
>
> "Never!" shouted the resounding voice of Daucus-Carota. . . . "Never . . . it will always be nine-fifteen. . . . The hand will remain where it was when time ceased to be, and your corporal punishment will be to observe the motionless hands and return to your seat to start anew, and this until you walk on the bones of your heels" [Ebin, 1961, pp. 14–15].

In Théophile Gautier's experience, time stops rather than disappears; eternity is duration without end, a state of perpetual suspense or impatience, punishment; eternity in hell rather than heaven, more or less as schizophrenics tend to describe it. It is not the timelessness of ecstasy, of the mystical, transcendental experience. There is an intimation of sadness in it, but also mockery, ambivalence rather than guilt or depression.

Marijuana in low or moderate experimental doses speeds up subjective duration, creating the impression that clock time slows down. In higher doses, marijuana creates a feeling of timelessness, leaving the subject uncertain as to the length of elapsed time. There is a confusion or disintegration of past, present, and future, which interferes with goal-directed thinking (Clark et al., 1970).

The temporal disintegration induced by marijuana may account for the experience of depersonalization frequently associated with its effects. Whether such a reaction is experienced as positive, negative, or mixed (ambivalent) seems to depend on the ego strength of the subject and his concern about losing control or suffering identity diffusion. Persons with a psychotic potential experience marijuana's fragmentation of the temporal sense and the concomitant discontinuity of the self with terror. The experience of temporal disintegration that characterizes marijuana intoxication may be caused, at least in part, by an impairment in immediate memory—William James's specious present. As Melges and his co-workers (1971) point out: "When an experience appears to have no before or after, it is isolated from time and, hence may seem timeless" (p. 209).

Time and Other Drugs

Even though distortions of time are typical of psychedelic or psychotomimetic drugs, any chemical substance can elicit such reactions. In an experiment with intravenous ethyl alcohol that I personally conducted (Hartocollis, 1962), one of my subjects said: "Time passed quickly, as it does when one has fun." And another: "Time didn't seem the same—longer or shorter, shorter I would say, time flew by" (p. 384). Other notable effects were feelings of depersonalization, a sense of irresponsibility, euphoria and gregariousness initially, followed by a state of relaxation, fatigue, sleep, or dejection. Subjects tested in groups displayed hypomanic behavior, a mixture of elation and boisterous, jocular hostility—all present-oriented affects or attitudes.

Alcohol accelerates the experience of external events, an effect that has been attributed to a slow-down on one's internal biological clock. Similar results—a shorter judgment of duration—have been obtained with barbiturates (Goldstone et al., 1958; Goldstone and Kirkham, 1968), nitrous oxide (Steinberg, 1955), heroin (Zinberg, 1971), and lowering of body temperature (Hoagland, 1933; Pfaff, 1968). Using the concept of the internal clock to explain the psychopharmacology of time judgments, Ehrensing and his co-workers (1970) have reported "a decreased event rate with sedatives and an increased event rate with stimulants, such as amphetamines and thyroid, and with increased body temperature" (p. 857). When the rate of internal events decreases, there is an increase in the duration of external clock time, the subject perceiving objective events as occurring at an increased rate and objective, clock time as passing quickly. Alcohol's effect in slowing one's hypothetical internal clock may account in part for the increased probability of having an accident while driving under its influence. As internal time slows down, the drunken driver would be likely to increase his driving speed in order to catch up with what seems to him accelerated external time—"to reach [his] destination 'in time,'" as Ehrensing and his co-workers put it (1970, p. 858).

Drug Abuse and Its Effect on Time

Danger—the threat of injury or loss, the preoccupation with an ominous or uncertain future—makes one feel that times goes fast. The feeling is one of anxiety, dread, fragmentation; and that is often worse than depression, or the worst aspect of a depression that is based on the notion of failure and a morbid preoccupation with the past that has not succeeded in eliminating the sense of time as future, the hope of survival or rescue of the self, a hope that can lead to real death by suicide. In such a case the motive of suicide is the search for timelessness, the wish to escape from the awareness of time as the bearer of bad news or the executioner of a cruel sentence. If time cannot be stopped, time has to be killed, and with it the self. Within limits, alcohol and drugs serve such a purpose. Characteristically, alcoholics talk about "oblivion" as the desired effect of their drinking, obviously meaning more than the elimination of painful memories or anxiety-provoking concerns, for they keep on drinking even after they manage to feel good. They actually mean to eliminate the sense of self that comes with the elimination of the sense of time. And they end up in stupor.

The use of marijuana and heroin by American soldiers during the Vietnam war was also for the explicit purpose of altering the experience of time. As Zinberg (1971) reported one veteran's discovery of heroin:

> The great shock came when [Tom] decided to go to bed. He thought perhaps a half hour had passed since he had smoked, but when he looked at his watch it was three and a half hours later! At first the realization was terrifying, but before he went to sleep he hugged himself with the delight of one who had discovered a secret weapon against time [p. 112].

Heroin "speeded up" time. As Zinberg's (1971) soldier put it: "It makes time go away. The days go bip, bip, bip" (p. 120). Marijuana "slowed up" time (presumably the awareness of time rather than the flow of time as such), providing the means to "getting through a thoroughly unpleasant year" (Zinberg and Weil, 1969). The goal was to reach a state of timelessness, the ability to experience the self beyond the bounds of time, as in the pursuit of the mystical experience.

Chronic use or addiction to drugs might be expected to make some difference, ordinarily in the direction of habituation, which requires a greater drug amount to produce the same degree of reaction. Paradoxically, the effects of marijuana, subjectively but also physiologically, are often milder or unnoticeable with nonusers, at least on their first exposure (Weil et al., 1968). But whether or not a particular drug reaction changes qualitatively with chronic use is highly questionable. At least so far as the effects on time are concerned, drug habituation seems to make little difference (Hoch et al., 1952).

Baudelaire's description of his experience with hashish differs from that of nonusers serving as subjects in the laboratory only in its poetic exaggeration. The testimony of another notorious opium addict, the English essayist DeQuincey (1822), sounds equally familiar:

> The sense of space, and in the end the sense of time, were both powerfully affected. Buildings, landscapes, etc., were exhibited in proportions so vast as the bodily eye is not fitted to receive. Space swelled, and was amplified to an extent of unutterable infinity. This, however, did not disturb me so much as the vast expansion of time. I sometimes seemed to have lived for seventy or one hundred years in one night; nay, sometimes had feelings representative of a millen-

> nium, passed in that time, or, however, of a duration far beyond the limits of any human experience [pp. 112–113].

DeQuincey also reports that opium intoxication brought about the rapid, hallucinatory recollection of early life events, some of which he had completely forgotten:

> The minutest incidents of childhood, or forgotten scenes of later years, were often revived. I could not be said to recollect them; for if I had been told of them when waking, I would not have been able to acknowledge them as parts of my past experience. But placed as they were before me, in dreams like intuitions, and clothed in all their evanescent circumstances and accompanying feelings, I *recognized* them instantaneously [p. 113].

Psychological Determinants of the Drug Experience

The dosage and purity of a particular drug, suggestion and environmental circumstances ("set" and "setting"), character traits and psychopathology are among the likely determinants of the drug experience. Personality, when described in global enough terms, has been found to be a factor in the direction of the drug effect. Thus, LSD is more likely to produce a "schizophreniform" reaction with subjects characterized by introversion, and more of a hypomanic reaction with extroverted ones (van Ree, 1960). A psychotically depressed woman who ordinarily complained that time appeared "absolutely endless," when placed on LSD reported that her time disorder was accentuated; she experienced time as an "awful state of eternity . . . [a] horror or eternity which I have had since I was a child and I just know it's something inside me which is wrong" (Kenna and Sedman, 1964, p. 283). Yet, when it comes to psychiatric illness, the drug effect is not always as predictable. In fact, schizophrenics have been described as "resistant" to LSD, in that they seem to react in a milder or neutral way when compared to normal subjects. This apparent tolerance of schizophrenic subjects may be related to the dose of the drug or to other factors such as the "macroscopic" nature of most studies that rely on statistical findings (Krus et al., 1963). And as they concern the experience of time, results may sound contradictory if the subject's personal account is taken at face value. A well-known source of confusion lies in the fact that the flow of time is described in diametrically opposite ways, depending on whether one is referring to it as an immediate experience or as a recollection.

In trying to account for the effects of LSD, Klee (1963) postulated a loss of the ego's "ability to employ defenses in the most adaptive manner" (p. 471). A more comprehensive explanation, similar to the one used by Freud (1900) with regard to dreams and by Rado (1933) with regard to alcohol, may be that psychotomimetic drugs like LSD, mescaline, or hashish suppress the vigilance of the superego and, as a consequence, some basic ego defenses such as repression, while enhancing other, more autonomous ego functions, in particular perception and the awareness of the present. The opposite obtains with obsessive-compulsive individuals who, under the influence of a strong superego and while sober, complain that time and the world around them go too fast for the way they feel inside, which is too slow or, in cases of depressed individuals, immobile.

Obviously a person's emotional experience has a great deal to do with the way he experiences the flow of time and vice versa, without necessarily the one being the cause of the other. It is more likely that the two components of the experience, its time sense and the pleasure-unpleasure dimension, balance each other; and in the extreme, the one dominates the other to the point of wiping it out. What accounts for the synthesis or decomposition (aposynthesis) of these two dimensions of experience is the integrity of the self, the degree of its vulnerability to stress, or more specifically the self-assessment of one's adequacy or inadequacy within a given perceptual field. Drugs seem to affect one's basic state of being by enhancing or diminishing one's instinctual drives, perhaps differentially with regard to their libidinal or aggressive nature.

A person's ego strength, as determined by innate factors and epigenetic development, beginning with self-object differentiation and object constancy, has a great deal to do with the self-assessment of his adequacy or inadequacy in coping with the neurochemical effects of the drug. The flow of time may be experienced as fast ("time flies"), but that could be either pleasant or unpleasant, depending on whether one's ego is under strong instinctual pressure, in particular pressure from libidinal drives. If enhanced by the effect of the drug, these drives create a state of anxiety that is characterized by the feeling that the self is immobilized while time is flying, the whole experience being unpleasant; if the instinctual drives are diminished, anesthetized, or somehow controlled by the drug, the self is experienced as flying along with time and time hardly experienced as such, the whole experience being described as a gratifying state of timelessness. On the other hand, when the immediate perception of time is experienced as slow, the overall experience is pleasant or unpleasant, depending on whether the ego is under the pressure of aggressive instinctual drives, directly from the id or via the superego. If the drug happens to enhance this aggressive energy, the individual has the feeling that his self is "on the run," experiencing an inner motion that contrasts with the perception of external time as slow and is felt on the whole as unpleasant. If, however, the drug weakens or somehow inhibits one's aggressive drives, the self is experienced as going along with time, and the resulting overall experience becomes that of pleasant timelessness.

Another way to look at the experience resulting from the use of drugs is in terms of their ability to influence one's perception of the future. As concern about the future—a superego function—diminishes or disappears under the drug's influence, the sense of duration as time that moves merges with the sense of self in a way that, as Huxley put it, makes the passage of time "indifferent." The flow of time becomes ego-syntonic and is translated into such expressions as "timelessness," "expanded time," or "eternity." The same experience, so far as the sense of time goes, when compared to clock time is likely to be described as fast, especially in retrospect.

X
TIME AND THE BRAIN

EVEN THOUGH THE BRAIN WITH ITS anatomical structure and neurochemical reactions underlies all thinking, feeling, and willing, nothing in its condition can really account for any particular thinking, feeling, or willing in which a person may engage. Human behavior is determined by personal motives that depend as much on the experience of the individual as on his constitutional endowment and actual environmental circumstances. Yet, in spite of the elusive, multidetermined causation of experience, scientists, neurophysiologists, and neuropsychologists in particular, have by no means abandoned the effort to identify the locus of mental functions in the brain, the search for specific mechanisms and "solid" analogues with which to describe the nature of mental phenomena or human experience, including the experience of time.

The Locus of Circadian Time

It may be generally assumed that psychological time, like language and rational, secondary-process thinking, is a function of the left or dominant cerebral hemisphere, while the perception of space is grounded in the right side of the brain. Such an assumption, implying that the perception of time is not as basic or primitive as the perception of space, would be in agreement with Piaget's developmental theory of time, while seemingly contradicting Kant's and Bergson's belief that the notion of time is at least as primary as that of space.

It should be recalled, however, that time as a psychological entity is not identical with conceptual time. Experiential time is an intuition akin to a primitive sense of efficacy or causality, antedating the development of conceptual thinking and the notion of time as measurable, object-related duration, or as self-centered perspective, time as present, as future, or as past. Furthermore, when considering the developmental origin of the experience of time and its localization in the brain, the key to the question lies probably deeper than the cerebral hemispheres. Indeed, recent animal research has lent considerable support to the idea that there is a biological internal clock system responsible for the circadian, sleep-wakefulness organismic rhythm.

The term *circadian,* derived from the Latin *circa dies,* denotes a biological periodicity that approximates twenty four hours. There are two kinds of circadian time: (1) an exogenous one, depending on environmental synchronizers, i.e., external periodic changes; and (2) an endogenous periodicity, an internal rhythm that depends on biological factors and, once established, functions like a self-sustaining clock, the environment merely modulating the frequency of its phases. According to recent laboratory findings, such a biological clock is located in or near the superchiasmatic nucleus of the hypothalamus, subject to the overall control of the pineal gland or epiphysis and its production of melatonin (Axelrod, 1974; Binkley et al., 1977).

Distortions of Time and Brain Dysfunction

Disturbances in the sense of time similar to those reported by psychiatric patients and people under the influence of drugs are known to occur during the prodromal phase of epileptic seizures. Dostoevski's (1868) account of the epileptic aura of Prince Muishkin in his novel *The Idiot* conveys a vivid and authoritative, if not typical, picture of the experience:

> He remembered that during his epileptic fits, or rather immediately preceding them, he had always experienced a moment or two when his whole heart, and mind, and body seemed to wake up to vigour and light; when he became filled with joy and hope, and all his anxieties seemed to be swept away forever; these moments were but presentiments, as it were, of the one final second (it was never more

> than a second) in which the fit came upon him. That second, of course, was inexpressible. . . . That there was, indeed, beauty and harmony in those abnormal moments, that they really contained the highest synthesis of life, he could not doubt, not even admit the possibility of doubt. He felt that they were not analogous to the fantastic and unreal dreams due to intoxication by hashish, opium or wine. Of that he could judge, when the attack was over. These instants were characterized—to define it in a word—by an intense quickening of the sense of personality. Since, in the last conscious moment preceding the attack, he could say to himself, with full understanding of his words: "I would give my whole life for this one instant," then doubtless to him it really was worth of lifetime. . . . The prince had confessed unreservedly to himself that the feeling of intensive beatitude in that clouded moment made the moment worth a lifetime. "I feel then," he said one day to Rogojin in Moscow, "I feel then as if I understood those amazing words—'There shall be no more time'" [pp. 213-215].

Disturbances in the perception of time have been correlated with electroencephalographic (EEG) activity, being observable during episodes of slow rhythm (Holubar and Machek, 1962; Anliker, 1963; Werboff, 1962). Such disturbances have been associated with damage in various areas of the brain. Thus, following bilateral surgery involving the dorsomedial thalamic nuclei, schizophrenic patients experienced a transient impairment in temporal orientation, being unable to identify correctly the date and hour of the day, their age and length of their stay in the hospital, or length of time since their operation. There was no consistent direction in their misjudgment, some patients overestimating, others underestimating time intervals (Spiegel et al., 1956). On the other hand, the case of a patient who lost the ability to experience the passage of time as a result of a right parieto-occipital lesion was cited by Gooddy (1969), who referred to similar findings by other investigators, notably Hoff and Potzl's observation that patients with lesions in the right occipital lobe felt that time was passing with "horrible rapidity." According to Gooddy, "when the two hemispheres are out of phase—as in the case of vascular or neoplastic lesions—a time lapse phenomenon may manifest itself" (p. 249). Similarly, the idea has been advanced that experiences involving an altered sense of time may be due to a disturbance in the transition of information from the right to the left side of the brain. As Efron (1963) put it:

> there is a fundamental mechanism or process related to the requirements of "time-labelling" of input and output signals which *is* located on the dominant temporal lobe and . . . damage to this area and its radiations may result in symptoms of disturbance to these "higher functions" . . . concerned with time analysis [p. 419].

Hughlings Jackson (1880, 1888) attributed the sense of familiarity or *déjà vu* and the associated phenomenon of "dreamy states" that characterize epileptic seizures such as Dostoevski's to lesions of the anterior and deep portions of the temporal lobe, bilaterally. The notion found experimental support in the work of Penfield (1959, who designated a certain area of the temporal lobe adjacent to the major speech area on the left and the space area on the right as the "interpretive cortex," for presumably rendering present experience meaningful.[1] Stimulation of this area, whether by a spontaneous epileptic discharge or by a surgeon's electrode, activates a record of experience somewhere in the grey matter of the higher brain stem or diencephalon. Forming the background of the stream of consciousness, it contains memories of life events that, trivial as they seem to be or entirely forgotten, may have a hidden meaning of particular significance to the individual (Kubie, 1954). Stimulation of the "interpretive cortex" causes these memories to return in "flashback" fashion, with remarkable sensory immediacy and with more detail than one can voluntarily recall. "The mechanism is capable of bringing back a strip of past experience in complete detail without any of the fanciful elaborations that occur in a man's dreaming" (Penfield, 1975, p. 34).

[1] Penfield points out that, since excision of this frontal lobe area does not do away with memory, the actual record of the past must be contained elsewhere, presumably in the diencephalon.

There is evidence that the area of the brain giving access to this record of consciousness is the hippocampus, the inner surface of the temporal lobe that forms part of the phylogenetically older limbic system—the so-called visceral or barbaric brain (Dimond, 1980)—and unites the activity of the temporal lobe with the mid-brain and the orbital surface of the frontal lobe (Williams, 1966). As part of a circuit that includes the "interpretive cortex," the hippocampus bilaterally monitors "experiential memory," while the limbic system of the brain stem provides a motivational mechanism or, as Ey (1969) put it, "a link between the integration of the experience of time in the normal field of consciousness and the desire to recall the past and anticipate the future" (p. 122).

Time Distortions in Near-Death Experiences

A combination of depersonalization and *déjà vu* phenomena involving a dramatic slowdown of the experience of time is often described by persons who have faced the specter of imminent death and sometimes survived. Dostoevski (1868), whose death sentence for his youthful revolutionary activities was reprieved only at the last moment, must have drawn from his own experience when writing about the distorted sense of time of a condemned man on the road to his execution:

> Then he is dressed and then begins the procession through the town to the scaffold. I think he, too, must feel that he has an age to live while they cart him along. Probably he thought, on the way, "Oh, I have a long, long time yet. Three streets of life yet! When we've passed this street there'll be that other one; and then that one where the baker's shop is on the right; and then shall we get there? It's ages, ages!" Around him are crowds shouting, yelling—ten thousand faces, twenty thousand eyes. All this has to be endured, and especially the thought: "Here are ten thousand men, and not one of them is going to be executed, and yet I am to die." Well, all this is preparatory.
>
> At the scaffold there is a ladder, and just there he burst

> into tears. . . at last he began to mount the steps. . . and meanwhile he notices and remembers everything [pp. 61–62].

Noyes and Kletti (1976) collected and analyzed over a hundred accounts of men and women with near-death experiences related to a variety of emergency situations, from accidents to suicidal attempts and critical illnesses, including cardiac arrest. What such people reported most frequently was an altered experience of time, in most if not all cases the feeling that time had expanded, slowed down, or stopped: "It seemed like the whole thing took forever," one subject said; "During all of this, time stood still," said another; and another: "My experiences were close to being in no time at all, almost as though time were at a standstill" (pp. 21–22).

Extraordinarily vivid or speeded up thoughts, a sense of detachment and a feeling of unreality were usual accompaniments of the experience. In a little over half the cases there was a striking lack of affect. When affect was reportedly present, it was described as one of extreme calm or ecstatic pleasure.

Sir John Eccles (Popper and Eccles, 1977), the Nobel Prize laureate physiologist, sees in such experiences evidence for the existence of a "self-conscious mind"—a force which, even though constantly and irretrievably grounded on the brain, becomes a free entity, interacting with the brain's neuronal processes so as to synchronize them with the circumstances of one's experience and adaptive requirements of the outside world. As Eccles puts it:

> When acute emergencies arise, time seems to run in slow motion. This must be an arrangement of the self-conscious mind reading out from the modules [of the brain] all of the acute input relating to the emergency. . . so that it apparently has more time to make decisions in the emergency. It has refined the experience of time into smaller pieces, you might say, for its actions, so that it has the best opportunity of countering the emergency [p. 529].

To illustrate his point, Eccles recalls a personal episode of near-death experience, worth reproducing for the vivid immediacy of its detail:

> . . . It was a very acute emergency when I thought I was going to be killed in a road crossing in Switzerland. We were doing a left turn into a main traffic road with nothing in sight, but it so happened that the low sun was in our eyes, and the road was heavily covered in trees. Down that dark road there was a dark red truck hurtling at perhaps 80 mph down the hill. My wife and I didn't see it until it flashed out of the darkness into the light. It was too late to stop so that all we could do was to try to accelerate to get out, and we were moving slowly because we had just come off a standing start! As I watched this truck coming closer and closer, time seemed to go on forever. I could watch it, thinking now I am past it, it won't hit me directly. We can get the front of the car out. It was getting closer and closer and then I thought it would hit the back of the car only, and then I thought, if it was the back of the car, we shall be spun around and perhaps crushed. Then in the end miraculously, I found that that the back of the car even wasn't hit and the truck moved past, but all in slow motion. It was the most incredible experience, and my wife had the same experience that time had almost come to a stop in this emergency. And so we drove on without ever daring to look back. The truck driver didn't seem to have seen us and made no attempt at braking. We had to do it all on our own.
>
> This was the self-conscious mind making in the emergency the very remarkable performance of steering and accelerating so as to get across [pp. 529–530].

As Noyes and Kletti have pointed out, the slowing down of time experienced during the brief period of critical danger seems to refer to the outer world, while the rate of inner time becomes correspondingly increased. Such a phenomenon is reported often by neurotic or psychotic patients and as a rule by people under the influence of drugs in the absence of any external emergency. On the other hand, it has been observed that the subjective slowing down of time characterizing a near-death experience is less likely to occur if the subject happens to be under the influence of alcohol or drugs, or to have a history of religiosity, possibly because such conditions render the idea of death less threatening (Greyson and Stevenson, 1980). In most cases the experience is characterized by a striking absence of affect or, rather, by a state of mind that is more spiritual than emotional. Like mystical or religious experiences, it is said to have an ineffable quality. When trying to describe it, the subject finds that his words fail to convey its vividness and importance.

A less common but most striking part of near-death experiences is the spontaneous and rapid unfolding of mental images representing old life events, especially happy scenes from early childhood, long forgotten. The phenomenon, which has been described especially in connection with accidents of falling or near drowning, may be triggered by a desperate attempt for reassurance (Greenson, 1971). But one cannot help wonder: what is its connection with the "flashback" experience triggered by Penfield's electrical stimulation of the "interpretive cortex" and by drugs like mescaline, heroin, or opium?

According to Schilder (1942), the *déjà vu* experience of epileptic patients stands for the "unconscious desire and fantasy of returning to the mother's womb" (p. 214). Hughlings Jackson (1880, 1888) had already noted that the *déjà vu* sense and "dreamy states" characterizing focal or temporal lobe seizures were similar to spontaneous experiences reported by personal acquaintances and literary figures such as Coleridge, Tennyson, and Dickens, who gave him no reason to suspect that they had a brain lesion. To illustrate the point, he quoted a passage from *David Copperfield:*

> We have all some experience of a feeling which comes over us occasionally, of what we are saying and doing having been said or done before, in a remote time—of our having been surrounded, dim ages ago, by the same faces, objects, and circumstances—of our knowing perfectly what will be said next, as if we suddenly remembered it.

Reflecting on the phenomenon of "reminiscences," Jackson (1880) reasoned that, rather than being aroused by a disordered part of the brain, it is "owing to but slightly raised activity of healthy nervous arrangements, consequent on 'loss of control'" (p. 313). In other words, *déjà vu* and related phenomena should not be considered as manifestations of pathology, but as defensive reactions to critical situations, near-death experiences or, in the case of brain damage, an epileptic discharge and its impending "loss of control."[2] This daring view, expressed by Jackson a century ago, fits Penfield's data, and foreshadows the psychoanalytic interpretation of phenomena involving an altered sense of time and the self.

Remarkably similar, disturbances of time and memory reported by psychiatric patients, drug users, epileptics, and people with near-death experiences remind one of the way time and the past are experienced in dreams. As in dreams, time is expressed in terms of space, as long or slow time being represented by slow movement in space and space that never ends, never comes closer in terms of its landmarks or destination or, better, the subject's destination in space. One may note in passing that destination, like fate, is a time-bound concept.

[2] It is curious that Freud would use Hughling Jackson's term "reminiscences" to describe the repressed memories of patient's suffering from conversion hysteria, memories recoverable by hypnotic suggestion or by free association in the model of a temporal seizure. But while Jackson reasoned that such reminiscences were defensive phenomena related to organic pathology elsewhere in the brain, Freud (Breuer and Freud, 1893–1895) proposed that these were related to a psychogenic conflict having to do with sexual wishes.

XI
TIME AND THE DREAM

In his conversations on Freud, psychoanalysis, and more specifically the idea of "a dream language," the Viennese philosopher Wittgenstein (1966) recalls the following experience:

> I was once looking at an exhibition of paintings by a young woman artist in Vienna. There was one painting of a bare room, like a cellar. Two men in top hats were sitting on chairs. Nothing else. And the title: "Besuch" ("Visit"). When I saw this I said at once "This is a dream," (My sister described the picture to Freud, and he said "Oh yes, that is quite a common dream"—connected with virginity). Note that the title is what clinches it as a dream—by which I do not mean that anything like this was dreamt by the painter while asleep. You would not say of *every* painting "This is a dream." And this does show that there is something like a dream language [p. 43].

Wittgenstein does not explain how its title—"Visit"—made him feel that the painting was a dream. Freud's alleged interpretation of the painting as being connected with virginity seems to be based on the assumption that the "visit" of the two men in top hats ("phallic symbols," as Wittgenstein is ready to acknowledge) takes place in a prostitution house. But why Freud agreed with Wittgenstein that the painting depicted a dream—actually, just the theme of a dream—we are not told either. I submit that the explanation is in the way the picture deals with time, the intimation of the experience of waiting for something in a bare room, in a place anywhere, anytime.

The Dreamer's Temporal Orientation

Psychoanalytic theory holds that, inasmuch as they are dominated by the unconscious, dreams are not bound by the constraints of time. According to Freud (1900), in dreams, as in psychosis, "there is a complete lack of sense of time" (p. 91). But as is the case with the unconscious, the timelessness of dreams is a relative matter. Both the experience and the abstract notion of time may be present in a dream, even though often in disguise. And when the notion of time is explicitly present, it means something different from what its manifest content indicates. The idea of timelessness for both the unconscious and the dream as an expression of the unconscious refers more to the immediacy of the experience and the seemingly illogical way time is dealt with than to the elimination of time as either experience or concept from the dream or from the unconscious as such. Above all, the idea of timelessness refers to the fact that the contents of the unconscious and of the dream have their origin in the remote past, representing childhood wishes and conflicts that, in spite of the passage of time, have lost none of their appeal or intensity. In the following typical dream, time is neither conceptualized nor experienced as such, but is implicit symbolically.

> A middle-aged man dreamed he was at a campsite with his teenage son, when high above the trees he noticed a large round basket, suspended in the air like a kite. Somehow he knew that in the basket there was a girl he had been in love with as a teenager, who had left him for another young man, not to be heard of again. She was often on his mind, and had been in his dreams repeatedly, especially in earlier years. Feeling an urgent need to meet her again, it occurred to him that he could reach her by invoking a genie locked in a bottle that his teenage son had in his possession in their tent nearby. His son asked the genie to take them up there, and soon both of them found themselves suspended in the air by a rope without visible end. Holding on to the rope tightly but effortlessly, he was soon at the same level with the basket, inside which he saw a woman, lying down, asleep.
>
> "Ruth," he said, "look at my son."
>
> She woke up immediately.
>
> "Larry!" she said, and paying no attention to his son, rushed forward and offered him her lips.
>
> He felt surprised at the ease with which she recognized him,

> for even though she looked as young and lovely as in the old days, she was somehow different from the way he remembered her. Reflecting that her having neglected him for so long gave him the right to be angry, he hesitated. To his dismay, she did not persist, but lay down and seemed to be going back to sleep.
>
> "Ruth," he called again, "look at my son." Then he began losing altitude and gradually fell to the ground. He looked around and noticed that the basket was moored at the far end of the campsite by a long rope. He thought he could use it to climb up again, but woke up.

In his dream, time and the past are represented by the rope that the genie provided for him and his son to reach the basket with his high-school sweetheart way up in the air. The timelessness of his wishes is represented by the girl's youthful appearance and eagerness to have him back. The fact that she looked different, even though as young and lovely, suggests that his wishes were even earlier than those associated with his teenage sweetheart, presumably the longing for reunion of a child with his mother. The reference to his son is another indication to that effect. It was as if he were saying to her: "look at your son. Look at me, I am your son." At the same time, the reference to his son was in recognition of the fact that he was no longer the young man she knew when he was his son's age—his son playing the role of a mirror reflecting himself through time.

It is also obvious that the dream involves an ardent sexual wish, which is frustrated, presumably by the woman's withdrawal at his hesitation. But both his hesitation and impotence may be ultimately related to the association of his old sweetheart with his oedipal or preoedipal mother. When this connection was interpreted to him, the patient confessed that his liaison with Ruth was never consummated because of his inability to maintain an erection the few times they tried to have sexual intercourse. At that time he blamed his inexperience; but, he added, knowing that the girl had had sexual relations with an older man long before they met, something that used to make him sad and sometimes angry, may have been the real cause of his impotence. He then agreed that the girl's first lover stood for his father.

In a dream everything is experienced much more immediately than in waking reality. Along with the fluidity that characterizes one's pursuits and beliefs, there is a sense of urgency or eagerness, a massive narcissistic involvement, an intense investment in the action, in the people, in the purpose that make up the dream. We explain it on the basis of the intensity of the wish that creates the need to dream, the need to experience an altered state of consciousness, one characterized by a broader perceptual field, a fluidity of self-boundaries.

As is the case with our wishes and affects, which regardless of their time orientation are always felt in the present, time as an experience in dreams is never attributed to anything but the present, even when the experienced events are based on affects and wishes of a different time. As Freud (1900) himself put it: "dreams make use of the present tense. . . and by the same right as day-dreams. The present tense is the one in which wishes are presented as fulfilled" (p. 535). His celebrated Irma dream—the "dream specimen of psychoanalysis" (Erikson, 1954)—was offered to us in the present tense. Here is the beginning of it, in Erikson's literal translation:

> A great hall—a number of guests, whom we are receiving—among them Irma, whom I immediately take aside, as though to answer her letter, and to reproach her for not yet accepting the "solution." I say to her, etc. [p. 10].

The dreamer's temporal orientation is strictly in the present. Any indication that the dream action represents childhood memories or wishes—that it is motivated by the wishes of the past—is under disguise, to be interpreted from the dreamer's associations. When there is open reference to the past, it may be assumed that there is a disguise in it as well.

We may dream of having feelings that we used to have in the past and, once awake and sufficiently removed from the impact of the dream, we no longer recognize in ourselves; however, it is rare to dream of ourselves as being of a different age.

We may possess a different identity in the dream, and that identity may be of a different age; yet the consciousness of age remains that of the present, at least implicitly. As Zilboorg (1951) put it:

> In the dream one could appear as a mature person (mostly in a dream), walking in the street and suddenly find oneself lying in a crib which the nurse is pushing along. While one is in that dream there is no surprise. There is no surprise while dreaming because there is no concept of opposites, no conception of time, no conception of continuity [p. 65].

Indeed, the experience of the dreamer is ageless, representing the subject's personality in its enduring traits, a self-concept that is related more to early wishes and identifications than to its current age and social role or professional identity.

Time in the Image of Space

In defending the notion of the unconscious, Freud (as reported in Nunberg and Federn, 1974) acknowledged:

> This does not hold absolutely true of the dream because it is a process of taking its course *between* consciousness and the unconscious. . . . The dimension of time is linked with acts of consciousness. . . [p. 291].

Elaborating, Federn pointed out:

> In remembering a dream we often do not know what was earlier and what was later: because a dream reaches consciousness, it must be perceived in the context of time, yet because of the timelessness of the unconscious the dream may perhaps have no orientation in time. It is true that in its deepest layers, the unconscious functions without the dimension of time [p. 304].

Upon listening to Federn's comments, Freud concluded that the latent content of the dream belongs to "a system in which the element of time plays no role whatsoever. On the other hand, this system does display something that resembles what we call 'spatiality' when we refer to objects" (p. 308).

On a number of other occasions Freud (1900, 1933) pointed out that time in the dream is expressed in terms of spatial relations, the manifest content of the dream transforming time into space. In Freud's (1933) own words:

> In general, indeed, where it is possible,the dream-work changes temporal relations into spatial ones and represents them as such. In a dream, for instance, one may see a scene between two people who look very small and a long way off, as though one were seeing them through the wrong end of a pair of opera-glasses. Here, both the smallness and the remoteness in space have the same significance: what is meant is remoteness in *time* and we are to understand that the scene is from the remote past [p. 26].

An elaboration of Freud's idea and specific example appeared in a dream of a young woman in analysis with a student of mine.

> I walked into a gift shop. There was a display counter and on it were some statuettes. As I walked by, something dropped by me. Then the irate store owner walked up to me and told me I would have to pay for it because it had been chipped. I picked it up and I told him, "No way this is chipped, I have one at home, and it is exactly like this." As I walked away, mad, I felt that he was following on my steps. I kept saying under my breath "Bastard, dirty son of a bitch." Then I noticed a funny device like a telescope that was stretching slowly—the strange thing was, I was looking through the wrong end, I was seeing the owner's face full of anger changing into my stepfather's face.

In associating to the dream, the patient recalled feeling excited, scared, and defiant when her stepfather beat her as a little girl. The appearance of the telescope in the dream, stretching slowly, brought back the memory of the stepfather's penis and angry face—distant, as if through the wrong end of the telescope.

Why space rather than time? What makes dream language prefer to deal in spatial rather than temporal terms, to represent the concept of time in terms of space? Space and time as intuitive notions of the mind, rather than as operational concepts (mathematical, clock and calendar time, or geometric space identifiable in terms of length and shape), are presumably primary. They define our perceptual reality and, whether they are structural givens of the mind or derive from coenesthetic experiences, they are not reducible to one another. The dream work reduces thought to its most intuitive elements, pictorial and linguistic forms that convey their meaning according to the three principles Freud identified as symbolism, condensation, and displacement. Spatial and temporal concepts follow the same dream rules.

Time, however, is not a simple notion. Time is experienced, and later on conceptualized, either as duration or as perspective. The experience of time as perspective—time as past or future—is predicated on the separation of self- from object representations, on the awareness of self-boundaries apart from the boundaries of objects—all of which implies the notion of space. But it is a different matter with the experience of time as duration, time as the intuition of being and not being, of feeling that things endure, that time moves along, slowly or fast, with or without the self. What determines the experience of time as duration is the instinctual pressure and the relative vitality of our wishes in the unconscious. In the dream, it is manifested in terms of motion rather than space as such, slow or fast movement that may refer to dream events as viewed by the dreamer—the dreamer being an observer—or to the dreamer himself as a subject in situations of entrapment, haste, or emergency, where movement or lack of movement is somehow time-bound, as in the following example:

> I was with someone riding in the night, a landscape with a bridge. It was a dark, eerie place, with dead trees along the road, dark, dark trees that were dying. It was dark, and water running high under the bridge. We were going to an antique shop to get a crystal bowl, like the one my daughter broke recently, and we had to cross the bridge. There was some desperation about doing that on time, we were waiting for someone, my husband, but you had to cross the bridge now, otherwise you wouldn't be able to do

> it again. We went ahead and crossed the bridge, and my husband didn't come.

Of course, motion implies space; and the dream just cited is remarkable for its landscape, which is recalled elaborately but as a background for action, as if designed to bring out the idea of change, of movement, of transition, of becoming now, in the present. The dream above conveys the urgency of a conflict which may have its roots in the past and may have repercussions for the future but is experienced as duration, as the process of decaying ("trees dying"), as emergency or internal pressure ("water running high under the bridge"), as the need to hurry ("you had to cross the bridge now. . . there was some desperation about doing that on time"), as the process of resolution ("we went ahead and crossed the bridge"), rather than as something in perspective, a childhood memory, an unresolved conflict from the past, a secret desire, or some concern about the future.

Time and Object Relations

There are dreams, however, in which the notion of time is quite explicit, as clock or calendar time (objective, mathematical, or abstract time), rather than time implied in images of space and motion (subjective time). It has long been known that the presence of objective time in dreams represents a variety of meanings, usually around past experience with time and its passage—anniversaries, deadlines, contracts, fate, death. Thus, according to Freud (1913):

> [The time of day in dreams] very often stands for the age of the dreamer at some particular period in his childhood. In one dream a quarter past five in the morning meant the age of five years and three months, which was significant, since that was the dreamer's age at the time of the birth of his younger brother [p. 194].

In another instance, again as reported by Freud (1913), time in terms of age represented emotionally charged moments of the past:

> A woman dreamt that she was walking with two little girls whose ages differed by fifteen months. . . . It occurred to her that both the children represented herself and that the dream was reminding her that the two traumatic events of her childhood were separated from each other by fifteen months (3½ and 4¾) [p. 195].

Dreams conceal one thing or another, defensively, for the purpose of protecting sleep. As Marie Bonaparte (1940) put it:

> This flight from the tyranny of time which we are able to review each night seems. . . to represent one of the greatest wish fulfillments accomplished by the dream for the benefit of human beings who remain harnessed to time's chariot while day lasts [pp. 461–462].

Presumably dreams conceal the awareness of time in order to escape the vigilance of the superego and the call to goal- or duty-bound reality. Thus time-related concerns in dreams may appear in disguise as figures involving transactions in terms of currency. For example, in a dream reported by Freud (1900, p. 414), a certain amount of money available to pay for "something" was found to represent the length of time the dreamer could afford to remain in treatment. Otherwise, when time is explicitly present, it stands for something different from what the manifest dream content indicates. Whether implicit or explicit, both kinds of reference to time stand for conflictual, emotionally charged object relations.

According to Gross (1949), abstract, objective time in dreams has reference to problematic relations in the present; concrete or alluded time to traumatic relations in the past. In my experience, and as the dreams quoted earlier from Freud's work may indicate, a good argument can be made for the opposite case: implicit reference to time in dreams may signify concern about a current relationship; while explicit, direct time may signify a conflictual relationship in the past or a traumatic event in the early family and childhood in general—the loss of a love object, sibling or oedipal competition, the threat of castration. Dreams of the latter kind tend to be repetitive, recurring periodically since childhood—another indication that they may refer to a past event or to a situation that has its roots in the past. During the analytic process, such dreams tend to become incorporated in the transference.

An obsessive-compulsive man, in analysis for about three years, had a series of dreams, all of which involved a strong sense of time:

> I had an odd dream last night, the patient recalled. It was as if I was an observer back in the fifties. It was in a small industrial town in Pennsylvania, people were on strike against an order by President Eisenhower, an order invoking the Taft-Hartley Law or something like it. I was supposed to be reporting on the strike and then, somehow, there was talk about a sex scandal, an affair involving a union leader and a widow, a woman in her late thirties or forties. She was complaining that the union leader was rude and had exploited her and so on.

When asked whether he had the awareness of being in the "fifties" while dreaming, the patient ascertained that that was indeed the case, adding that it was like seeing an old newsreel. Then he recalled watching a television show about the McCarthy era the previous night. He also recalled that after the show he had the wish to be there, to stand up against McCarthy, feeling that Eisenhower had acted "very shabbily." At this point the patient became restless and declared he had the sense the dream was not important, it was like a crossword puzzle that had no meaning really, like an intellectual game he was playing by being concerned about such a dream. "I don't know why I brought it up," he said, "it is not connected with me, I have no real feeling for it. I barely remember it, there is such a distance from it, like something that happened during the night and I don't have the sense of any continuity with my real life. I was acting as an observer, I was not terribly interested in the first place, I didn't know where my loyalties lay, there was something unsavory about the way that man treated the poor woman." I pointed out that it was paradoxical to call the dream unimportant and not related to his concerns when he described its content in some detail and expressed intense feelings about it. Reluctantly, he acknowledged that the dream had more meaning than he wanted to believe. He felt attracted to "shabby" love affairs, and he thought he might himself want to behave in the way the labor union leader did in the dream. "Somehow," he said, "I identified with the old man. And yet all the persons involved in the dream were emotionally quite distant from me. I was a reporter doing a story on either side, neither of them was too appealing, etc." At the end of this new disclaimer of feelings, he confessed that he was currently interested in a young woman, a friend of his wife's, whom he found sexually exciting.

The case illustrates a number of points. To the extent that the reference to time in the dream's manifest content and experience is oblique or vague (a period described in terms of a historical figure or event) it pertains to a current interpersonal conflict—the temptation of an extramarital affair. The reference to the past serves as a disguise for the fact that the problem is current. To the extent that it specifies an age or date, the dream refers to an old conflict, oedipal or early adolescent competition with father, as suggested by the fact that the date in the dream coincided with the patient's late childhood. The patient's awareness of the particular time is that of an observer, someone who is emotionally detached from the latent content of the dream. A day later, the same patient reported the following dream:

> I was in a strange city, wandering around this one area. It was like a golf course, and I found some balls lying around. I walked through a swampy area and it turned out to be the middle of town. The dream had something to do with trains, with schedules, going to some other city, traveling around. What stood out in the dream was walking up some place to be on time, some kind of schedule, something to do with trains, being on a train. This one was more like living it, not stopping to observe what I was doing. There was more action than anything else, I was so engrossed in doing things that I can't tell what I was doing. There was a background of interest, enjoyment I suppose, feeling good afterward.

In this dream the reference to time is indirect, contained in the movements of the dreamer through space, the concern with train schedules, and the sense of traveling. Even though the reported experience is pleasant, this may be more in its secondary elaboration than in the actual dream experience, which carries a certain degree of anxiety implicit in the dreamer's preoccupation with schedules and the vagueness of his destination. The patient did not come up with any meaningful associations from childhood. He kept returning to current concerns about his family and the guilt he felt for not spending more time with his wife, his frustration with his employment, his heavy work schedule and preoccupation, which did not allow him to finish a paper he was writing for some time now or to concentrate better in his analysis, where he had usually the feeling that he was "meandering" aimlessly, except for yesterday when he felt he had discovered something meaningful, which made him happy for a while. But now he could not tell what that was. Some time later, the same patient reported a third dream:

> I was in the middle of some war activities, a World War II battle. I was with a General, standing on top of a hill overlooking a valley where there was another army, the enemy, under another General, who was supposed to be MacArthur. I had the feeling that my General was very indecisive, very slow. The time went by and nothing happened. It was like a calendar in front of me—the days, the months, the years went by, one by one, without much happening.

The patient brought up a number of associations related to experiences of the weekend and his feeling that he was not engaged enough in life; his wife did not want to entertain guests, preferring to stay home with him alone; he felt awkward with his professional engagements, and so on. I reminded him of the fact that just the previous day he was complaining about the slowness of his analysis and blamed the analyst for not being alert enough or interpretive in the right way, wishing to have a different analyst, one who had impressed him with his sense of organization and knowledge of theory. The patient acknowledged his concern about his slow progress, but thought that it was not entirely the analyst's fault. He recalled that as an adolescent and later, until he started in analysis, he had the fantasy of being General MacArthur's lieutenant and somehow proving to be as decisive a leader as MacArthur. And, as before, he recalled feeling impatient with his father as a child, having the feeling that his father was just a showman, without inner strength or real effectiveness.

Noteworthy also in this dream is the use of spatial concepts and images (standing on top of a hill overlooking a valley) to suggest the sense of time as perspective—surveying the past, re-experiencing the past from the distance of the present. In both instances, however, time is represented objectively (as the idea of past, future, clock time, calendar time) as well as subjectively, i.e. implicitly (as in the idea of distance or in a situation involving frustration, haste, an emergency of sorts).

Indirect references to time in terms of situations of frustration or emergency usually involve the immediate experience of time as duration. Such an experience is often related to conflicts involving the superego, object relations generating intense ambivalence and guilt. Inasmuch as the idea of time remains implicit, such relations are likely to have reference to current conflicts—to conflicts involving contemporary relationships. The experience of time as short (or fast) may signify the wishful search for a love object that is either prohibited, or unattainable. That was the case in the second dream (p. 157), where the dreamer felt in a hurry to cross a bridge threatened by rising water. She was dissatisfied with her marriage, entertaining the fantasy of running away with a more glamorous man than her husband, ideally with her analyst, toward whom she nurtured strong erotic feelings at that point in her analysis.

Time may be experienced as short (or fast), either explictly or in situations of frustration involving a missed appointment, a departed train, and so on. When time is experienced as long (or slow), either explicitly or in situations of frustration involving waiting and obstacles of one sort or another, the dreamer may signify guilt over a love object that is ambivalently held. It may be in response to the aggressive part of the ambivalence that the dream, defensively, keeps the object away, out of reach, to preserve it.

Time and Affects in the Dream

Freud (1900, and as reported in Nunberg and Federn, 1974) held that dreams involving the idea of departing on a journey and missing the train are motivated by the wish to deny the fear of dying. Referring to Stekel's notion of departure as a symbol of death, he pointed out that the dreamer wishes to reassure himself by saying, "you are not going to die, you are not riding along [on the train]" (Nunberg and Federn, 1974, p. 219). Such an interpretation implies an unconscious reference to time as future. To the extent, however, that the fear of dying derives from the childhood fear of castration or separation from a love object, the true perspective of the dream is with reference to the past. Freud, indeed, compared the missed-train dream to another common recurrent dream, the so-called examination dream, which handles anxiety over some current problem by restaging a problematic situation from the past, a classroom examination, that one has been able to overcome.

A special sense of time that refers to a strange familiarity with a certain landscape or locality, what is known as a *déjà vu* experience, when it occurs in a dream has, according to Freud (1900), a special meaning: "These places are invariably the genitals of the dreamer's mother; there is indeed no other place about which one can assert with such conviction that one has been there once before" (p. 399).

To the extent that affects, beyond the elementary affective polarity of pleasure-unpleasure, are time-bound and dreams have an interest in neutralizing disturbing affects (typically by displacement of their ideational representation to an indifferent object or event [Freud, 1900]), one might expect them to become disengaged from the dimension of time. As I have shown elsewhere (Chapter V), certain painful affects, such as anxiety, are predicated on the idea of time as future, contained in the idea of confrontation with some vague or explicit danger to the integrity of the self; while another category of painful affects, such as depression, are predicated on the idea that something detrimental to the self (a loss or failure important enough to lower one's self-esteem) has happened and lies, irrevocably, in the never-land of the past. The elimination or displacement of its component sense of time as future or past would either neutralize the painful affect or convert it into a present-oriented affect, such as anger or boredom (restlessness, impatience), with which a dream can cope by means of immediate action. The following example may illustrate the point:

> My father was murdered by some robbers; they held my mom, and she had to watch and they cut him into pieces. They cut his legs and arms off and he died praying. The pieces were nailed to a board. The first couple of days I behaved like nothing happened. Two days after he died it hit me, and I cried and cried at the funeral.

The dreamer felt nothing while her father was dying in the process of being mutilated and crucified (incidentally, the father was a minister). Devoid of time perspective, his death elicited no affect in his daughter and allowed her sleep to continue. The introduction of the time element as past ("two days after he died") made the father's demise meaningful, and the dreamer experienced appropriate feelings of sadness and guilt. How she was able to witness the father's tortured death without horror (a feeling dominated by the sense of time as present) or action may be explained by assuming that the idea of mutilation was displaced from the father's genital organs to his extremities, and that the blame for the cruel act was projected onto the mother, who was made to watch the murder in the dream.

Ordinarily, time can be felt directly as the sense of duration, time moving slowly, fast, or not at all; or indirectly, as tense or perspective, in the context of some time-bound affect, such as anxiety with its sense of expectancy or anticipation of something dangerous for the integrity of the self. But when the affect becomes too intense, the sense of time as perspective becomes secondary or disappears and is replaced instead by the sense of time as duration; until affect and time blend into one ineffable, unbearable experience, and the person loses consciousness. When that happens in a dream, the person wakes up. This is usually the case with nightmares, which contain both an element of fear or pain and the experience of time as dragging out or standing still, a time sense that is the contrary to the experience of timelessness. Erotic dreams have a similar composition and fate: thick, intense pleasure combined with the feeling that time is running out, as if compressed in a capsule or going so far as to disappear into nowhere. Paradoxically, as the person wakes up from such dreams, he feels as if they have lasted an eternity; and he has a hard time shaking off the affect.

Dreams involving a massive discharge of affect stem in all probability from early childhood experiences, referring to the primary object relations with mother or father. They tend to recur. They are in essence screen memories of early object relations, as in the following recurrent dream (borrowed from an article by Smith [1977, p. 315]).

> I am in this space—a nothingness—with no darkness but a diffuse light as if I am on an infinite plane in a kind of foreverness. In the dream [last night] it was like walking through air where everything is white, but the air has some kind of resistance to it like gelatin. There is a corollary dream of being with my mother in some sort of space but with a sense of rooms and soft carpeting—but all with a colorlessness, whiteness. My mother is only a shape or a presence, and I am very small. It's like getting away to someplace where everything is sufficient.

There is here the suggestion of a longing for a close relationship similar to the one that presumably exists in an early phase of infant development with mother, a symbiotic relationship that is characterized by a spaceless quality (a "nothingness") and timelessness (a kind of "foreverness") that is neither spaceless nor timeless really, but embedded in a world of infinite space and time, a boundless world that is described by adults as a mystical or religious experience—an ineffable experience, as they would point out, because it sounds so contradictory when described in terms of space and time. The same contradiction is evident in the above dream. As mystics do, the dreamer is struggling to get around the paradox of representing space and time as senseless or nonexistent when he feels so positive about them, by invoking the sensory dimensions of sight and touch. He describes space in terms of a "diffuse light" and a "colorlessness" that is "really whiteness," and the experience of time as "walking through air [. . . that] has some kind of resistance to it like gelatin." He also identifies mother in terms of space and time as "only a shape or a presence," and the regressive act of reliving the experience in the dream, the reversal of time, as traveling through space away from the here and now, "getting away to some place where everything is sufficient." The awareness of the self as something "very small" next to the mother's "shape or presence" suggests the remoteness of the experience in time through the visual perspective of appearing small in the distance. The idea of smallness also suggests the beginning of self-object differentiation, something that is further implied in the identification of the space that mother occupies as one "with a sense of rooms and soft carpeting."

In fact, a dream is the closest thing to a mystical experience that an ordinary person may claim. The blissful experience we can gain in a dream seems to be predicated on the ability of the self to become oblivious of all sensory input, to either ignore it or dissociate it from its material, objective source in the outside world or within one's own body. Whether a metaphor or psychophysical reality, the idea that the body ego is decathected and all available instinctual energy is concentrated on the mental images of the internal self-object world describes best, so far as I am concerned, the situation that prevails in the dream. Energy-like attention, withdrawn from sensory organs and voluntary muscles, cathects the visceral part of the organism, accounting for the warm relaxation, a melting-away feeling, experienced as we are about to fall asleep. When this energylike attention flows, so to speak, away from the mental representation of the viscera and cathects exclusively the mental representations of the internal self-object world, a dream is born: consciousness of the mental representations without awareness of the material, time-bound, objective world.

XII
AGGRESSION, MYSTICISM, AND THE SENSE OF TIMELESSNESS

UNLESS THERE ARE NO WISHES to gratify, time will be experienced as moving—moving faster as one's primary, instinctual drives and derivative wishes become weaker, moving slowly as they gain in strength and urgency. If all wishes, including superego expectations and demands, are overcome, time as an experience ceases to exist: one then speaks of timelessness, as mystics and others who claim to have had the experience describe it.

Varieties of Mystical Experience

The so-called mystical experience is an altered state of consciousness described variously as joyful, blissful, luminous, spaceless, and timeless. It can occur spontaneously or as a result of systematic preparation that may or may not include the use of drugs. Depending on the precipitating circumstances, it goes by different names. Thus, when elicited by prayer and ascetic exercises, it is identified as religious or mystical; when induced by hallucinogenic drugs, as psychedelic; and when occurring spontaneously, as cosmic illumination, transcendental or conversion experience.

The term mystical derives from the Greek *mysticon* (with its Sanskrit root *mus*), meaning "secret," the hidden. The term is related to the mysteries of the antiquity, the secret practices of the Dionysian and Orphic cults in ancient Greece. Combining aggressive and sexual acts, these led to orgiastic group behavior and an altered state of consciousness that was interpreted as supernatural, or as evidence of communication with superhuman powers. The use of artificial stimulants—such as the chewing of ivy leaves, loud music, and special lighting effects—was then, and has often been, part of such practices. The highly emotional nature of these experiences and the concomitant frenzied activity distinguishes them from the mystical consciousness that is connoted by the term "religious experience." What these two major types of mystical experience have in common is a heightened emotionality and unconcern with rationality. The mystical experience known as religious or psychedelic is characterized by a passive, quiet, and motionless, receptive state of being and a sense of peace, in contrast to the orgiastic mystical experience, which is characterized by motion and excitement. For this reason they have been also described as "introvertive" and "extrovertive," respectively (Stace, 1967). Both experiences seem to have in common an element of love; but while in the one love is spiritual and peaceful, in the other it is sensual and aggressive.

The Search for Unity and the Eternal

The mystical element in terms of the emotional, and as contrasted with the rational, is a universal aspect of the human experience. It is identified with the divine and the inspirational; or, as Plato put it, with the prophetic, the ritual, the poetic, and the exotic. The term mystical is related to, if not interchangeable with, the notion of suggestibility, introversion and intuitive knowledge—again, as contrasted with the rational, the analytical and the explicit: the evident versus the obscure; the apparent versus the hidden; the immanent versus the transcendental.

Furthermore, the mystical experience is characterized by an unshakable conviction, usually absent from the consciousness of everyday life. This is the sense of conviction that one finds in delusional states, spontaneous or drug-induced; the condition of being in love; and some very vivid dreams. But in the case of mystical experience, the person who claims to have experienced it feels no need to explain it or to defend it, accepting it as self-evident or self-sufficient. As Russell (1929) put it: "The mystic insight begins with the sense of a mystery unveiled, or a hidden wisdom now suddenly become certain beyond the possibility of a doubt. The sense of certainty and revelation comes earlier than any definite belief" (p. 9).

A fundamental characteristic of mysticism is its belief in unity, the assertion that reality is one and indivisible. There is no opposition or division anywhere; dichotomies, polarities, conflicts are fused, transcended or resolved; good and evil are one or, rather, the reality of evil is only apparent—reality is all good, and good is one. The mystic strives for unity or wholeness, for homeostasis, for the unchangeable or the eternal; and this he finds in the mystical experience.

According to the Upanishads of India, written about 2500 years ago, the mystical experience is "the pure, unitary consciousness where an awareness of the world's multiplicity has been completely obliterated. It is One without a Second." And according to Plotinus, the Syrian mystic who wrote in Greek during the third century, "In this seeing, we neither distinguish, nor are there two. The beholder has become the unity, having no diversity either in relation to himself or anything else." Finally, according to Martin Buber's contemporary testimony, "From my own unforgettable experience, I know well that there is a state in which... we experience an undivided unity... this basic unity of my own soul... is beyond the reach of all the multiplicity it has hitherto received from life" (cited by Stace, 1967, p. 27).

Adopting the concept of "de-automatization" that Gill and Brenman (1959) developed to account for the regressive process in hypnosis, Deikman (1966) suggested that the mystical experience of oneness is the result of the undoing of automatization of the ego and, in particular, its perceptual-cognitive apparatus. Fischer (1971), on the other hand, making use of evidence from experiments with drugs, described the mystical experience as the end result of a process of arousal of the subcortical nervous system, which the self on a cortical level is likely to interpret in stepwise fashion as creative, psychotic, or ecstatic—what Fischer calls "the perception-hallucination" continuum. When the subcortical system becomes aroused, the perceptual-behavioral apparatus or cortical level of function regresses to a more dependent status, losing its freedom to interpret the sensory data of the outside world as distinct from the self and inner experience. Hence the experience of union with God or the eternal universe.

The yearning for the experience of undifferentiated unity has been interpreted by psychoanalysts as a regressive wish for a preverbal state of being that is either objectless—a condition of "limitless narcissism" characterized by an "oceanic"[1] feeling (Freud, 1930)—or the self fused with the image of the nurturing object's breast (Lewin, 1950). According to Jacobson (1964), a similar wish—the wish for oneness with mother—is a normal phenomenon in the developmental history of the infant's mental life. As she put it:

> . . . for some years a child still feels himself to be only an extension of the mother and participates in her imagined omnipotence, or, the reverse, regards the mother as part of his own omnipotent self. . . . The child's resistance to give up such magic fantasies is, of course, expressive of his desire to evade the painful intrusion of reality by maintaining or by re-establishing his lost unity with his earliest love object, the powerful mother. . . . Evidently, such experiences of merging with the love object are always connected with a temporary weakening of the function of perception—i.e., of the awakening sense of reality—and with a return to the earlier, undifferentiated state [pp. 241–242].

According to Lewin (1952), mother in general comes to represent the pleasure principle, primary-process thinking, and timelessness.

> Poets and artists depict time as a father figure, but they see immortality in the sense of timelessness which is a gift of the mother, who in the beginning made us immortal by magical potions of sleep-bringing nourishment. Immortality or timelessness is an oral gift [Lewin, 1952, pp. 309–310].

[1]Homer in *The Iliad,* Book XIV, calls Ocean "the origin of the gods" and "the origin of everything."

The Fear of One's Own Aggressive Potential

Stimulated by a testimonial letter from an American physician seeking to convince him about the existence of God and after-life, Freud (1928) suggested that some experiences of religious conversion may be caused by the intense reactivation of oedipal feelings, and in particular aggressive wishes against the father. In agreement with Freud, Salzman (1953) explained conversion experience as a method of solving the conflict arising from feelings of fear and hate toward one's father, or for that matter anyone who might represent authority, including the mother.

A similar idea underlies Simon and Simon's (1972) attempt to account for Bertrand Russell's contention that his change from an imperialist young man with a violent temper into a lifelong pacifist was caused by a mystical experience. He was 28 and living with the Whiteheads when, returning home from a "deeply moving" recital of Euripides' *Hippolytus,*[2] he found his friend's and mentor's wife in the middle of a severe bout of chest pain. As Russell (1967) himself recalls in his autobiography:

> She seemed cut off from everyone and everything by walls of agony, and the sense of the solitude of each human soul suddenly overwhelmed me. Ever since my marriage, my emotional life had been calm and superficial. I had forgotten all the deeper issues, and had been content with flippant cleverness. Suddenly the ground seemed to give way beneath me, and I found myself in quite another region. Within five minutes I went through some such reflections as the following: the loneliness of the human soul is unendurable; nothing can penetrate except the highest intensity of the sort of love that religious teachers have preached; whatever does not spring from this motive is harmful, or at least useless; it follows that war is wrong, that a public school education is abominable, that the use of force is to be deprecated, and that in human relations one should penetrate to the core of loneliness in each person and speak to that [p. 220].

[2]The story of a young man who is falsely accused of having an affair with his stepmother.

The experience, which Russell describes as "a sort of mystic illumination," turned him into "a completely different person."

> Having for years cared only for exactness and analysis, I found myself filled with semi-mystical feelings about beauty, with an intense interest in children, and with a desire almost as profound as that of the Buddha to find some philosophy which should make human life endurable. A strange excitement possessed me, containing intense pain but also some element of triumph through the fact that I could dominate pain, and make it, as I thought, a gateway to wisdom. The mystic inside which I then imagined to possess has largely faded, and the habit of analysis has reasserted itself. But something of what I thought I saw in that moment has remained always with me, causing my attitude during the first way, my interest in children, my indifference to minor misfortunes, and a certain emotional tone in all my human relations [p. 221].

James (1902) described the process of religious conversion as an escape from the notion of sin, a notion reflecting one's aggressive potential and problems arising from the awareness of such a potential, problems experienced as anger, fear, worry, and despair. Characteristically, he cited as examples the cases of people like Tolstoy, whose religious conversion was preceded by a period of severe depression. Boehme, "one of the most original and astonishing of the Christian mystics" (Underhill, 1926, p. 214), described the feelings that led to his conversion as follows: "I fell into a very deep melancholy and heavy sadness, when I beheld and contemplated the great deep of this world" (p. 214).

As Salzman (1953) pointed out in discussing his own clinical experience with such people: "Hatred, resentment, and hostile, destructive attitudes seem in each case to have been involved in the preconversion experience" (p. 183).

Boisen's (1936) recollection of his personal experience with the phenomenon gives a vivid impression of the aggressive energy involved: "Strange and mysterious forces of evil of which before I had not had the slightest suspicion were also revealed. I was terrified beyond measure" (p. 3).

Commenting more generally on the proverbial wish of mystics to get rid of the notion of time, Arlow (Panel, 1972) has suggested that the wish corresponds to unconscious fantasies of escape from the pressure of their aggressive drives. Evidence from the life histories and psychological evaluation of several young patients who sought a mystical experience in religious movements or in the use of psychedelic drugs seems to support the idea that a yearning for such an experience stems from an intolerance of one's own inner violence, specifically, one's dormant hatred toward "bad" internal objects, beginning with the oedipal father and, conceivably, going back to the image of the frustrating or "exciting" mother of early infancy (Fairbairn, 1952). The following are illustrative cases from the C. F. Menninger Memorial Hospital.

> John B., a twenty-five-year-old university student, was in psychotherapy for a little over a year when he experienced dissatisfaction with the progress of his treatment, and requested that his therapist prescribe LSD for him. He felt that psychotherapy did not work because of his "built-in sabotage" efforts that kept him from knowing what was going on inside him. Referring to himself as "crazy" and to psychotherapy as "patchwork," he said he could live with his craziness but that he needed more than patchwork. He felt "LSD treatment" could help him "dig deeper," help him "stop bugging" himself in order to have peace of mind. He mentioned using methedrine when much younger and experiencing intense emotion while he was under the influence of this drug, especially coming to realize how angry and how violent he could become.
>
> He then recalled some years back thinking about his potential for violence, and a grave concern he had over it and a fear of what would happen unless he was continually aware of it. He remembered being "wild" as a child, and how his parents were always after him, trying to control him; how at the age of nine or ten it dawned on him that he was indeed uncontrollable and that, unless he resorted to what he called "a conscious reaction formation," he would likely be always this way. His violence was a problem he could not see anyone else in the world helping him with.
>
> Several hours later, Mr. B. told of his growing concern that

rumors were being spread about his being a homosexual. He spoke with anxiety and anger about the people who allegedly spread such rumors, but at the same time he described feelings of compassion and love for these people, now feeling enraged at them and the next moment loving them "completely." In fact, he believed that his inner violence was the result of a pent-up frustration over the inability to express this "complete love," that if he could express this love, the violence would be gone from his mind. And he feared that he might split apart or explode if he did not get over the barrier that kept him from expressing his love. To this end, he once more pleaded with the therapist to help him have "LSD treatment," promising to obtain the drug himself if the therapist agreed to supervise the session.

Dick R., a high-school drop-out, was seventeen years old when he was admitted to this hospital complaining that his inner conflicts interfered with his being as he wished to be. His parents complained about his impulsive behavior, disinterest in school, drug use, hippie-type friends, and their own lack of control over him.

As his mother recalled, the patient was preoccupied with violence since an early age. "My husband and I built a train table for Dicky when he was five, and from time to time we'd see to it that it was put together, like at Christmas time, into a pretty peaceful setting with houses and lights in the windows, little cars, railroad crossings, churches, police station, cows in the pasture. It was a battle to keep it that way until after Christmas, when the wars always broke out, with the whole table covered with soldiers painted with red paint for blood, some with legs cut off, not a house, not a church standing, and the table would continue like this in an endless battle until the following Christmas." Both parents thought that the patient had always a "low tolerance for doing anything specifically asked," recalling that on one occasion, at the age of five, he became so angry with his father that he kicked him in the shins. On another occasion, at thirteen, when his father attempted to discipline him, they got into a fist-fight that left both of them hurt, the father with a permanent eye injury.

The boy's school performance and relations at home began deteriorating rapidly when he was in the seventh grade. He expressed the feeling that his mother hated him and wanted to poison him; and he abused her verbally to the point that she became quite frightened of him. He began experimenting with LSD and, dropping school, he moved into a commune, where he came under the influence of an ex-priest, began meditating, and wrote poetry. But soon he felt depressed and let his parents arrange a psychiatric examination.

Even though the patient talked about love and peace, his examiner experienced him as a sullen, hostile, angry young man. Psychological testing suggested he was prone to impulsive outbursts, which he experienced as alien and frightening. His anger was more often controlled and expressed in impersonal negativism and a hypercritical attitude toward the world, which was seen as a place where people are exploited, robbed, oppressed, and forced to conform or be destroyed.

After six weeks of formal psychotherapy, his therapist reported: "The patient has been more agitated and expresses concern over his ability to control his own violence, a capacity which he no longer denies. His ideology of love and peace seems to be at odds with his own life experience, and he finds himself questioning the need for such ideology, with a resultant anxiety. . . . He is slowly finding himself unable to maintain his steadfast rebellion against all values which he believes his father supports."

Martha V., a high-school drop-out, was seventeen years old when admitted because of feelings of depression and anxiety, which developed after she became involved in a fundamentalist religious organization. Her parents described the patient's early behavior as "aggressive" and "persistent." During the first grade, she underwent psychological testing, which showed "clear-cut indications of anxiety related to aggressive behavior and some inability to deal with thoughts related to bad behavior." From the second grade on, however, the patient was a straight-A student, and the teachers described her as a sweet child, who did what was expected of her.

At the age of fifteen, upon her return from a vacation trip in Europe, Martha felt suddenly bored and reluctant to go back to school. Shortly afterward, a girl friend persuaded her to attend a meeting of the Catholic Pentecostal Group, and at this time she had her first of a series of religious experiences. She went through a spell of happiness, but then she experienced anxious and depressed feelings, which soon developed into social withdrawal, loss of interest in school work, and insomnia. She reported recurrent nightmares or hallucinations, consisting of Satan coming after her in the form of a cat, or sitting in her room in the form of a chair, or being a big, tall person in the process of molesting her. After a suicide attempt, she was taken to a local psychiatric hospital, where she impulsively attempted suicide twice more before transferring to this hospital.

Psychological testing showed that the patient was "aware of dangerous, aggressive impulses, which she tried to suppress and control by appearing benign, harmless and vulnerable. Acceptance of the reality of conflict, disharmony, and aggression in the world around her appeared to be a serious threat to her own control over hostility. The patient's struggle with her aggressive impulses seemed to dominate her psychic life, with little room for age-appropriate interpersonal relationships. In addition to controlling aggression, she was struggling with infantile oral wishes of support, nurturance, and supply."

Distortions of Time and the Wish for Selflessness

As suggested in Chapter VIII, people who complain that time has stopped are struggling against dangerous sexual or murderous wishes about which they feel intensely guilty. To believe that time has slowed down or stopped is reassuring to them in that the fulfillment of their guilt-ridden intent, or what they expect to be their own destruction, may be postponed indefinitely.

The idea that the appreciation of duration as slow or fast is a function of one's motivational state and, more specifically, of the closeness of instinctual, libidinal or aggressive wishes to consciousness is implied in experimental findings such as those reported by Deikman (1963). During sessions of meditation conducted according to specific instructions from the mystics' literature, his subjects felt that time had passed very quickly, or that less time had elapsed than what was actually recorded on the clock.

Dooley (1941) has suggested that an obsessive concern with the loss of control over the sense of time reflects the fear of loss of "the world of objects" and, therefore, of the integrity of the self. On the other hand, Bergler and Róheim (1946), writing about the regressive wish to get rid of time, pointed out that the sense of timelessness is evocative of a lost state of blissful, objectless existence, reexperienced as a "fantasy in which mother and child are endlessly united" (p. 103).

One who is not concerned with time is likely to experience oneself as happy—timelessness or the sense of eternity being identified with the condition of ecstasy that religious mystics describe in their writings. I am quoting from Eckhart (as cited by O'Brien, 1964):

> . . . for there exists [for God] only the present instant. But this man abides in that same light with God. That is why there is in him neither suffering nor succession [of events], but an eternity ever the same. This man, in truth, is able to be astonished at nothing; in him all things are in their [eternal] essence. Thus, is he the recipient of nothing new nor of any accident whatsoever, for he abides in a Now which, always and without end, is itself new [pp. 152–153].

Religious mystics describe their goal as the achievement of a state of selflessness, which they equate with perfection. Using ego-psychological terms, Fingarette (1958) compared the mystic notion of selflessness to a normal experience which is "non-anxious and motivated by neutralized drives functioning within the nonconflictual portions of the ego. It is an unselfconsciousness akin to the normal awareness of our breathing" (p. 38). It is selflessness in the sense that the ego is purged away from anything that is conflictual and divisive, whatever is aggressive or sensual and makes the wish for the "good" object greedy, hateful and anxiety-provoking, dangerous to the object world and to the self. About a thousand years ago, Saint Symeon the New Theologian (1792), one of the founders of the Greek Orthodox ascetic movement of Hesychasts surviving until recently on Mt. Athos, wrote a set of instructions that are suggestive of the same concern with one's inner aggression and greed in wishing to reach a mystical experience, in this case by means of the so-called *little prayer of Jesus*.

> Three things you need to guard above all. First, freedom from concern about anything reasonable or unreason-

able and vain, that is, mortification of everything; second, clear conscience about everything... and third, complete freedom from effort without wish for anything material. Then sit down alone in a corner at a covered and quiet place and close the door and collect your mind away from everything temporary and vain; put your chin on your chest, so that you can watch within yourself through the eyes of your mind and through the eyes of your senses, and hold a little your breath so that you can concentrate better, and search with your mind to find the place where your heart is, so that your mind settles altogether there. And at the beginning you are bound to find there darkness, great depravity and cruelty. Later on, as you engage in the work of the prayer incessantly day and night, you will find—like a miracle!—an endless joy; because, as it struggles for it, the mind will succeed to find the place of the heart, and then immediately it sees in there all those things that neither saw before nor knew. For it sees that air which is to be found in the heart and sees its entire self bright and full of every wisdom and distinction. And from then on your mind chases away the evil thoughts with the prayer: "Lord Jesus Christ have mercy on me" [pp. 86–87].

When Aggression Prevails

In describing the regressive process inherent in mystical experiences, Jacobson (1964) pointed out that an excess of aggressive potential may lead to a psychotic breakdown:

> . . . normal experiences of ecstasy do not aim at destruction but are founded on a fantasy of libidinal union between self and object world, [resulting] in a transitory sense of self-expression and the feeling that the self and the world are rich. Such experiences of merging, which may briefly re-transform the images of the self and the object world into a fantasy unit vested with libidinal forces, permit an immediate re-establishment of the boundaries between them. By contrast, pathological regressive fusions caused by severe aggression may result in an irreparable breakdown of these boundaries and hence of the self and object representations [p. 69].

In their wish to rid themselves of all violent potential and the concomitant threat to the internalized "good" object images or, for that matter, threat to the self, people whose early object relations were dominated by aggression may fall into a delusional state and risk personality disintegration rather than finding inner unity and the condition of selflessness described by religious mystics. The suspension of rational, secondary-process thinking and the lifting of ego defenses involved in the search for a mystical experience tend to reactivate an early stage of inner object relations in which the aggressive part of the self overwhelms, "swallows" so to speak, the libidinally invested part; or, more specifically, the "good" self-image is "swallowed" by the "bad" mother image. Thus, instead of achieving the desired blissful condition of oneness with the "good" mother, the self runs the risk of becoming lost in a chaotic world of dreadful and hateful "bad" or "split" internal objects. The prevalence of aggressive potential would explain the experience of fragmentation described by clinicians like Searles (1965) as being characteristic of the schizophrenic condition. And the "bad trip" that so many young people experience while experimenting with psychedelic drugs may attest to the same dynamic relation between a heightened awareness of one's inner violence and wish to get rid of it. The following cases, borrowed from a clinical report by Rosen and Hoffman (1962), may help illustrate the point further.

> A young man in his late teens had been on a four-day LSD "trip" when he was forced into a homosexual relationship. As a result, he became convinced he was going to die. Feeling that he had offended God, he remembered the biblical verse, "And if thy right eye causeth thee to stumble, pluck it out and cast it from thy head." And he plucked out his right eye—"in order to rid my body of the devil," as he reported later.

> After dropping out of college, a young woman joined a hippie community in the country. One day, while hitchhiking, she was raped and became pregnant. After having an abortion, she returned to the hippie community, where she began taking LSD, becoming preoccupied with mysticism and the occult. Feeling increasingly guilty, she suddenly became assaultive and was taken

> to a hospital. There, in the belief that one of the nurses was working for the devil, and in the wish to save the world, she took her right eye out. She also quoted the same verse from the Bible.

Both patients in the above-cited cases had been the victims of sexual assault, after which they began taking LSD, searching for a mystical experience. But, instead of the desired freedom from violence in a world of love and selfless unity, they found themselves flooded by their own potential for violence; and, feeling that the devil was inside them, they tried to exorcise him by taking out one eye—the "evil eye" standing for the aggressive part of the self that threatens to destroy the image of the "good" mother and frustrates the wish to relive the ecstatic experience of the undifferentiated all "good" self-object state of early infancy, the wish to reunite with the all "good" mother in a timeless and spaceless world of peace and love.

The ability to eliminate the sense of time, whether through the use of drugs or in the course of a mystical experience, brings about a state of being that is free from the awareness of its boundaries; in other words, it eliminates the sense of space as defined by the self and its boundaries, but without destroying them. And that is what makes for the ineffability of the experience. There is no meaningful way to communicate one's experience in terms of ordinary, secondary-process logic, or to conceptualize the self without the sense of time and space. Under the circumstances, selflessness, or the absence of the sense of self, has been described as nothingness: "It may fitly be compared to Nothing," as the mystic Jacob Boehme put it; or, according to Angelus Silesius, "God is pure Nothingness, neither Now nor Here can touch him" (cited by Sterba, 1968).

Man's native aggression becomes object-bound and, under optimal circumstances, neutralized in the ego, while remaining relatively raw and mobile in the mental structure of the superego, providing it with its threatening or punitive function experienced as guilt or remorse, respectively. It is significant for the role that the fear of one's own aggressive potential plays in mysticism and the wish for the timelessness of the mystical experience that the superego would be centrally involved in man's ability to conceptualize time, to convert the primitive experiences of duration and perspective into orderly, predictable clock and calendar time.

It is obvious that the essence of the mystical experience, as described by a score of ancient, medieval, and contemporary experts, is in its complete freedom from sensual or aggressive wishes, the elimination of all internal instinctual pressure, and the perception of the external world as devoid of any exciting elements, threatening or tempting possibilities. The concept of time as a dimension of reality that defines self from object is canceled and replaced by a sense of unity, which suggests a return to the putative, preambivalent symbiotic state of being and blissful "oceanic" feeling of early infancy, before the differentiation of self from object, a state of being characterized by Eckhart as "no willing, no need, no act, no generation, [but] an eternal silent undifferentiated One-ness" (in O'Brien, 1964, p. 149).

The remarkable thing about such a condition is that the person who experiences it does not lose consciousness of either himself or the surrounding reality. On the contrary, he is allegedly able to have a broader sense of reality concerning himself as well as the world around him. The consciousness of self as separate from objects is retained, while the consciousness of the self as the subject of wishes or fears and of objects as sources of pleasure or pain is lost. The self-image seems to take possession of the object world, or to lose itself in it without giving up its autonomy or destroying the integrity of the object image—a situation that, comparable as it may be experientially, is certainly not identical with any known phase of the infant-mother relationship.

XIII
MYSTICISM AND THE TIMELESSNESS OF THE ID: THE CASE OF NIKOS KAZANTZAKIS

As the violence of the world around him was reaching its climax, and barely a year before his own surrender to the violence of cancer and old age, Freud (1938) entered these seemingly irrelevant thoughts in his notebook: "Mysticism is the obscure self-perception of the realm outside the ego, of the id" (p. 300).

Freud was certainly not a mystic, not even in the sense that Einstein was (Lubin, 1977), but his interest in mysticism and the occult is well documented (Jones, 1953–1957; Zilboorg, 1967). That he was concerned with mysticism at this particular moment in his life should not be surprising either, for mysticism is associated with the yearning for peace and love that men experience when besieged by violence and death.

Mysticism and Violence

People wish for peace when they are either afraid or tired of war. But mystics and mystical movements do not usually appear in times of all-out war, when violence and privation are upon everybody. Mysticism seems to be a manifestation of personal conflict in times of peace, elicited by one's own inner aggression rather than by the violence of the outside world. The presence of external aggression, if directly menacing the individual, is not likely to produce a mystical reaction but, rather, fear and flight, or violence in self-defense. The mystical urge for peace and love develops in the context of a nominally peaceful society that promotes and vicariously enjoys violence, offering personal immunity from violence, but leaving the impression that it is futile to oppose it. Such conflictual external circumstances, reflecting a similar internal conflict, are likely to promote the wish for a mystical experience, which may provide peace and love for the individual concerned apart from the society in which he originates. To paraphrase Freud, mysticism is the obscure perception of the self in the timeless realm of the id.

The search for peace and love through mysticism, motivated by the wish to escape one's own potential for violence, does not necessarily lead to a peaceful, conflict-free mystical experience. In fact, quite often the effort to induce the regressive conditions necessary for a mystical experience results in the flooding of one's ego with violent wishes, bringing about a psychotic episode, a "demonic" rather than "religious" experience or, if the effort involves drugs, a "bad trip" rather than the desired "psychedelic" effect. I discussed this dangerous possibility in the previous chapter, illustrating it with a number of characteristic clinical examples. In the present chapter I shall try to show how a common psychosomatic syndrome, identified in the literature of mysticism with the struggle of ascetics against sexual needs and temptation, can be explained better as a manifestation of the mystic's struggle against aggressive wishes and the fear of violence. In support of this hypothesis I shall present the case of Nikos Kazantzakis, the modern Greek poet and novelist whose life and writings were guided by a mystical search for love, peace, and God as a universal, timeless principle. In so doing, I will point to the timeless quality of both sexuality and aggression or, more specifically, the reversibility of time when it comes to the experience of sexual temptation and the fear of losing control of one's own repressed aggression.

Kazantzakis' Illness: Its Sudden Onset

Kazantzakis was about to turn forty when he found himself stranded in post World War I Vienna, alone, suffering from a painless but disfiguring swelling of the face, which resisted all treatment efforts until he finally decided to get out of Vienna. From the description he himself provided, his affliction can be readily identified as an atopic or idiopathic neurodermatitis, commonly attributed to allergy. But, allowing for the possibility of allergic factors, there is enough evidence to suggest that Kazantzakis' condition was determined by aggressive wishes and fears of retaliation he harbored in his unconscious mind since early childhood. Kept under tight control by obsessive-compulsive defenses, an ascetic style of life, and a messianic ideology, such wishes and fears were precipitously reactivated by the casual encounter with a young woman in the semi-darkness of a crowded Viennese movie theater.

In his autobiographical book *Report to Greco,* which he wrote shortly before his death, Kazantzakis (1961) tells us that the occasion of this illness brought him in contact with psychoanalysis and convinced him that the power of the unconscious was stronger than his willful determination to become a religious prophet, adopting and propagating a Buddhistlike philosophy of life. Kazantzakis' account, presented in the chapter "Vienna, My Illness," refers to events that transpired about 35 years earlier and, besides being subject to distortions of memory,[1] were probably manipulated for literary purposes. But a written document of the same period, Kazantzakis' (1958) own correspondence with his first wife Galatea, offers a certain corrective to such reservations concerning the accuracy of the *Report to Greco.* Covering the entire Vienna episode from May to August 1922, it gives the history of his illness in chronological sequence and considerable detail.

The first letter in the series to mention this illness, written about ten days after Kazantzakis' arrival in Vienna, begins as follows:

> Chérie, if you suddenly opened the door of my room and saw me, you would feel much sympathy for me. I'm lying

[1] According to his translator and close associate Friar (1960), "Kazantzakis had an extremely bad memory for time, place, or enumerations" (p. 138).

> in bed, the face wrapped up in compresses that I change every half hour. I don't know where I caught [sic] eczema! You don't know what I'm suffering. . . . Alone here in the hotel . . . I fix with ineptness that you would see with much tenderness bandages, antiseptics, etc. I can't eat, or go out. But all this I endure with calm and patience [p. 24].

In a letter written on July 10, six weeks after the onset of the same illness, Kazantzakis informs his wife he is still not well:

> My illness comes, goes and comes again. The doctors don't understand anything. Only one, a great professor here, told me that it is due to psychological reasons. I will tell you when you come, because it is something curious [p. 48].

What Kazantzakis promises to tell his wife, but so far as it is known he never did, is revealed in the *Report to Greco*. The latter account attributes the onset of his facial swelling to an evening encounter in a movie theater. As Kazantzakis (1961) recalls:

> Next to me sat a girl. Her breath smelled of cinnamon. I felt her bosom heave as she respired. From time to time her knee touched mine. I shuddered but did not draw away. She turned and glanced at me for an instant, and in the half-darkness of the auditorium I thought I saw her smile [p. 353].

Before the show was over, he felt bored and got up to leave. The girl did the same. They walked together toward a park. He asked for her name. "Frieda," she replied. And then, laughing: "Why ask? My name is 'Woman.'"

Hardly realizing what he was doing, Kazantzakis invited her to spend the night with him. She said she could not, but promised to see him the next day. Whereupon, with a sense of relief, Kazantzakis left to return home. Suddenly, he became aware that his lips, cheeks, and forehead had begun to swell.

> My eyes soon grew so small that nothing remained but two slits, and it was only with difficulty that I managed to see anything at all. . . . When I finally arrived [home] and turned on the light and looked, I emitted a cry of terror. My entire

> face was swollen and horribly disfigured., My eyes were barely visible between two overflowing masses of florid flesh, and my mouth had become an oblong slot incapable of opening [Kazantzakis, 1961, pp. 353–354].

The Verdict of an Analyst

Kazantzakis hoped he would soon be well so that he might get together with the female encounter of his night at the movies. But in spite of all the help he received from internists and dermatologists, the swelling and disfigurement did not go away. Finally, suspecting that the cause of his affliction was psychological, he went to see psychoanalyst Wilhelm Stekel.

Referring to him as a "famous disciple of Freud," Kazantzakis explains that he was familiar with Stekel's work—something easy to understand, as Stekel happened to be not only an enthusiastic apologist of Freud's ideas, but also a regular contributor to the medical columns of the contemporary Viennese newspapers (Stekel, 1950). Kazantzakis told Stekel he had been wondering about the meaning of the world since he was a teenager. And, finding Christian religion naïve and overoptimistic, he had exchanged it for the moral tenets of Buddhism.

As Kazantzakis recalls, Stekel interrupted him: "To search in order to find the world's beginning and end is a disease," he told him. "The normal person lives, struggles, experiences joy and sorrow, gets married, has children, and does not waste his time in asking whence, whither, and why. But you did not finish your story. You are still hiding something from me. Confess everything" (Kazantzakis, 1961, p. 355). Kazantzakis then told him about Frieda and their agreement to spend a night together—at which point Stekel interrupted him again to tell him that he suffered from what was once known as "the ascetic's disease." He reminded him of the ancient hermit who, pressed by his sexual needs, hurried to the nearest town to find a woman, when suddenly, as he was about to enter the town, he looked at himself and realized with horror that his body was covered with leprosy. It was not leprosy, however, but the very illness that made

Kazantzakis unable to keep his date with Frieda. The hermit ran back to his desert, thanking God for having kept him from sinning; whereupon God forgave him and his symptoms cleared away. Likewise, pointed out Stekel, Kazantzakis' illness kept him from transgressing his ascetic principles. The only way to get well would be to get out of Vienna.

There are two more reports about Kazantzakis' illness in Vienna, attributed to the author himself but recorded by others. The Greek-American poet and editor Kimon Friar (1960) recalls that on two occasions, while working together on the English translation of Kazantzakis' epic poem *The Odyssey,* more than 30 years after the Vienna episode, he observed that the author's "lower lip swelled lugubriously to great size and dripped a yellowish fluid" (p. 8). It was during one of these occasions that Friar heard from Kazantzakis about his illness in Vienna, allegedly the first person to be told about it after Galatea and Stekel. According to Friar, Kazantzakis delegated to him the responsibility of revealing the story when the opportunity came for the sake of science, as a "neo-ascetic case of great interest" (p. 8).[2]

Friar's report is almost identical to a third account, given by the French writer and publisher Jouvenel (1959) on the basis of an interview with Kazantzakis a few months before the poet's death. These second-hand accounts differ from the one given by Kazantzakis in his autobiography mainly in the description of the circumstances of his encounter with Stekel. While in the *Report to Greco* it was the author who took the initiative in seeking out the popular and versatile psychoanalyst at his office, according to the other two accounts it was Stekel himself who, attracted by the stranger's monstrous appearance, approached him during a night at the Opera.[3]

[2]Kazantzakis' face did not swell again until the winter of 1938, and then a little over ten years later (July 1948) as he was writing about the swelling of Manolios' face in *The Greek Passion* (1948). Similar crises kept recurring with greater frequency until his death in the fall of 1957, 35 years after the initial episode in Vienna (H. Kazantzakis, 1968).

[3]Interestingly, both Friar and Jouvenel published their reports before Kazantzakis' *Report to Greco* appeared posthumously in print, and presumably neither one knew of its existence. Friar's account of Kazantzakis' alleged meeting at the Opera

Kazantzakis' Relationship with Galatea

The version that Kazantzakis gave in the last days of his life is more dramatic and in some respects more detailed than the story emerging from the letters he wrote to his first wife Galatea. Yet these letters contain material that he either did not mention to Stekel, or Stekel did not pay attention to, or perhaps Kazantzakis himself could no longer remember 30 years later. Kazantzakis was still married to Galatea and, in writing to her, tried to convince her to come and live together in Vienna; but, using one or another excuse, she procrastinated. Again and again he professed great love for her, and began to sound almost desperate as she not only refused to come, but teased him, doubting the sincerity of his feelings and the seriousness of his intentions, if not his cause itself. One might expect him to get angry and criticize her in return but, on the contrary, he kept praising her as "the most honest, bold and balanced soul" in Greece (Kazantzakis, 1958, p. 17).

In his letters from Vienna, Kazantzakis conveys the im-

---

offers some colorful details which, difficult as they may be to take literally, do remind one of Stekel's forceful, iconoclastic personality, as well as Kazantzakis' flambouyant, impressionistic style as a storyteller. According to Friar (1960), during a night at the Opera, "he was approached gingerly by a stranger who inquired if he might be permitted to ask a question. When Kazantzakis found himself acquiescing, the stranger politely asked, 'Would you mind telling me what role eroticism plays in your life?' Kazantzakis was taken aback and shocked, but the stranger hastened to explain that his question was purely clinical, that he was a professor of psychology at the University, and that he would look upon it as a great favor if Kazantzakis would honor him with a visit to his office the next morning. He then apologized once more and gave his name: Dr. Wilhelm Stekel" (p. 8).

According to the same account, Stekel received his Greek patient several times in his office, supplied him with literature (presumably pertinent to his condition, if not Stekel's own writings), and told him he was "planning to quote, in a future work, some remarks of [Kazantzakis'] about *eros*, and to describe three or four of his dreams" (Friar, 1960, p. 5). Remarkably, vague as all the other reference to Stekel is in Kazantzakis' correspondence to Galatea during the time of the Vienna episode, Kazantzakis does mention that his dreams had impressed "the Psychology Professor," as he repeatedly identifies him in the letters, and that the "Professor" planned to use them as well as the case itself in one of his next books. The letters, however, make no mention of the opera, or of the circumstances under which Kazantzakis made Stekel's acquaintance. Actually, according to the letters, the only time Kazantzakis went out in the evening during the period of his illness was to listen to a lecture by a "famous theosophist" named Steiner (Kazantzakis, 1958, p. 25).

pression of a man not only in love, but greatly dependent on his wife. He describes himself as lost without her, helpless and unhappy, thinking about her and pleading with her to come and join him. And he constantly asks her to do things for him, to transact his business with publishers, bureaucrats, and politicians, to find and send him books, money, clothes, and various household items. Actually, most of his letters read like a list of wishes and chores a child may entrust to his mother as she goes out for a shopping trip.

Paradoxically, even though lonely, unhappy, and insecure without Galatea, pleading with her to join his voluntary exile, constantly daydreaming about settling in some quiet place with her, he refused to stay with his wife for any length of time, being always on the move, seeking new departures, new beginnings, new knowledge, new experiences, and new exiles. Galatea later figured out that during the 14 years of their marriage, they lived together only four years (G. Kazantzakis, 1957).

In her autobiographical novel *Men and Supermen* (G. Kazantzakis, 1957), Kazantzakis' first wife describes their relationship as devoid of sexual interest. According to her, Kazantzakis tried to explain his passive, inhibited behavior toward her—what she interpreted as "hopeless narcissism" on his part—as caused by a "secret power" that stood between them and prevented him from behaving like a man. She remembers him saying: "As it happens with common folk, when the new groom cannot sleep with his bride because they have him bound by a magic spell, that's the way it happened with me; it was as if they had me bound with regard to you" (p. 208).

Almost 30 years later, writing about Jesus' last wishful dream from the Cross, Kazantzakis (1955) echoes the same idea of a prohibitive force that kept him from uniting with the woman he loved:

> Jesus rushed forward and took her in his arms. "Magdalene, beloved Magdalene" he cried, "oh, how many, how very many years I've longed for this moment! Who stepped between us and refused to leave us free—God?" [pp. 448–449]

Similarly, in the semi-fictional *Report to Greco,* there is a brief exchange with a prostitute in Vienna's Prater—what sounds like a preliminary version of his encounter with Frieda—where Kazantzakis excuses his aversion to sexuality by invoking the image of a Godlike figure, Buddha:

> "Come with me. . . ." She whispered, winking her eye.
>
> "No! No!" I cried as though in danger.
>
> She released my arm. "Why not?" she asked.
>
> "I'm sorry but I don't have time."
>
> "Are you crazy?" said the girl, glancing at me with sympathy. "What are you, a monk? No one is looking."
>
> "Buddha is looking," I was about to reply. . . . Feeling as though I had escaped a great danger, I returned post-haste to my room [p. 349].

Kazantzakis and His Father

Kazantzakis' excuse, "I don't have time," brings to mind the symbolic association between the notion of time and father (Father Time) that, according to psychoanalytic literature, marks the early development of the superego—"I don't have father's permission." Indeed, describing his fateful encounter with Frieda—the "Woman," as Kazantzakis calls her at this point, a liberated and liberating woman, as the choice of the name Frieda may indicate—he makes a curious, and probably significant, reference to father: the passing thought that his invitation to her is spoken with the voice of one of his ancestors, meaning his father and the wish to identify with father, a wish that is immediately denied as inappropriate. Reasoning that his father "despised" the kind of women Frieda represented, Kazantzakis tried hard to avoid the implication that she stood for the mother of his childhood—a forbidden object so long as he was not strong enough to eliminate from his mind the oedipal father, the formidable Capetan Michalis of his later novel *Freedom or Death,* or, for that matter, Buddha, the holy man he tried to reincarnate while in Vienna.

Kazantzakis' relationship with his father may be described

as intensely ambivalent, a blend of dutiful love, unmitigated fear, and barely contained hatred. Kazantzakis had little patience for his father's parochial ideas, but he was awed by the old man's quiet strength and sober "forbidding" demeanor, being careful not to provoke his wrath or disappoint his high expectations of him as the only son and heir. A dream he recorded in one of his notebooks a year or two after the Vienna episode is suggestive of his unremitted conflict with the oedipal father, pointing to an unresolved identity problem:

> There was a war going on, and the General looked at me and said, "Dont' go away! Come and fight too!" A sort of terror passed over me. . . . The General thought a bit and resumed: "But you can't fight on the right, because you look to the left. Nor can you fight on the left, because you also look to the right! So you will have to carry ammunition to both right and left, to both camps [H. Kazantzakis, 1968, p. 103].

Kazantzakis, as well as his second wife, who reported it, assumed that the dream manifested his mystical inclination to do justice to both sides of an argument, to recognize both sides of the eternal conflict that splits the human self and the object world into spirit and matter, light and darkness, good and evil —the wish "to balance. . . the frightening conflicting forces of the universe" [H. Kazantzakis, 1968, p. 114] in order to create a harmonious self, to elevate his human nature into that of perfect God.

Besides serving both camps, however, the act of carrying ammunition to both sides betrays the wish to keep them fighting with each other, to feed the conflict, to keep conflict alive. Actually, what excited Kazantzakis was conflict itself, which he saw constantly both within himself and outside as eternal, and which he tried to describe and glorify rather than solve in his novels, in his tragedies, and above all in his *opus magnum, The Odyssey.* According to his own account a few years before his death:

> The major and almost only theme of all my work is the struggle of man with "God": the unyielding, inextinguishable struggle of the naked worm called "man" against the

> terrifying power and darkness of the forces within him and around him [H. Kazantzakis, 1968, p. 507].

That the General of the dreams stands for his father becomes clearly evident in a related paragraph, to be found in one of Kazantzakis' letters to his first wife from Vienna. As he is trying to describe his wish for a new, more acceptable identity, he writes:

> I am the Lord your God, cries aloud inside me his screaming Voice. I am not shelter. I am not home and hope. I do not cure. I have not mercy. I am not good, I am not Father nor Son. I am Your General! [Kazantzakis, 1958, p. 223].

In the same way, the dream expresses his conflictual wish to identify with father as a man, an identification that is not possible so long as mother has not been renounced as an oedipal object, with all the concomitant terror of father's punishment by castration. But the dream also betrays the timeless wish to submit to father as a woman, the regressive wish to be a woman so as not to fear father's wrath but, rather, enjoy father's protection, contemptuous yet loyal and reassuring.

Kazantzakis described his father as a "grave, somber, wild man" (H. Kazantzakis, 1968, p. 207). According to the personal testimony of fellow Cretan and early schoolmate Fanourakis (1958), however, the poet's father was parsimonious with words, very serious and brave, but kind and merciful. It is known that he drank to excess and in the last years of his life had become very irritable and misanthropic. Following the death of his wife, a few months before his own end, he locked himself in his bedroom and wanted only to drink and to feed the birds that came to his window. He died in a state of confusion, probably from hepatic failure. In his delirious talk, he was heard to welcome his younger son, who had died accidentally as an infant while the family was trying to escape a Turkish pogrom. The fact that on his deathbed Kazantzakis' father hallucinated the return of his long-dead child seems to indicate that he experienced little pleasure at the thought of his elder son, who not only did not succeed him as a "warrior chieftain," but did not even manage to become a university professor,[4] as he had promised in the youthful letters he wrote to his father while engaged in postgraduate studies in Paris.

The news of his father's death found Kazantzakis once more abroad, this time in Spain. Ten years after the Vienna episode, he was now approaching fifty. In a letter to his fellow Cretan and literary disciple Prevelakis (1965), Kazantzakis describes his reaction as follows:

> My dear brother, I received your telegram. I left immediately, seized by inexplicable terror. I had the need to make my body tired, exhausted, so as to feel some relief. It was not love, something deeper, more animal-like, as if a large piece of my body collapsed suddenly on the ground, as if I was pulled by horrible roots into the ground, before my time. I wandered, Salamanca—Valladolid—Burgos—Zarugoza—Valencia—Alicante—Elche, 2.000 Kilom., without eating almost anything, without sleeping, with exhausting nightmares. . . . [p. 351].

In trying to explain his neurodermatitis, Kazantzakis claims that Stekel attributed it to a conflict over sexuality, to a determined struggle on Kazantzakis' part to suppress strong sexual wishes, in no way indicating that aggressive wishes or the fear of violence, his own and that of his father, were considered as part of the explanation. And yet, in writing about the psychodynamics of religious ascetics, Stekel (1950) suggested that they suffer from an obsessive-compulsive neurosis caused not only by repressed sexual drives, but primarily by the fear of death, their conflict with God and nature representing the struggle of social and spiritual ideals against the instincts of sexuality and death.

Throughout his life Kazantzakis was preoccupied with the idea of violence and the fear of death. Though an avowed humanist and pacifist, he was always troubled by destructive wishes, as is particularly evident in his letters to his first wife.

[4]In 1933, Kazantzakis applied for the Chair of Ancient Greek at the University of Madrid (E. Alexiou, 1966).

Thus, in describing the turbulent political situation and the economic depression that prevailed during the post-World War I period in Vienna, he wrote:

> Yesterday, there was a woman sitting on the edge of the pavement with one leg over the other and underneath you could see a horrible gray slip and her whole body, up to the navel, stark naked. She looked sad and cynical and most pale from hunger. . . . "Modesty" is a luxury for the rich. Ah! how well I understood this "sister" of our new religion. Better that the earth should be destroyed! That the firmament be purified of the shame of modern life! . . . My God, I cry aloud inside myself as I walk along the great roads, when will you descend like a vehement wind, like Megas descending from the peaks of Parnassus, to clean the earth! [Kazantzakis, 1958, p. 19].

What characterizes Kazantzakis' state of mind in Vienna is, on the one hand, great compassion for the plight of women and the weak; and, on the other hand, rage and a revengeful wish to destroy the old, oppressive and decadent order of things, what he identifies with capitalism and homosexuality—the oedipal father and his own regressive inclination to submit to his father's powerful, timeless will. He entertains fantasies of violence. He anticipates with excitement the proletarian revolution. He quotes with admiration the bloodthirsty words of local communist leaders. But his wish for violence is a conflictual one, full of fear and guilt. When he wishes for the revolution, he does it as a spectator. Wishing to see its catastrophic, purifying results, he is hardly ready to actively participate in it.

Kazantzakis grew up in an environment of violence. Crete, his native island, had been under foreign domination for centuries and was still struggling to gain its freedom at the time he was born. As a little boy, he lived almost constantly in a state of alarm, reinforced by his father's efforts to build into his consciousness the taste of violent death and the will to defend the honor of the family. In his *Report to Greco,* Kazantzakis (1961) recalls with awe the time his father lifted him in his hands and forced him to kiss the feet of some native rebels hung by the

Turks in the public square of their neighborhood; and another time, when his father, sharpening a long black-handled knife, had extracted from him the promise to kill mother and sisters if the Turks ever threatened to take them into slavery.

His violent but repressed disposition and its relation to the father of his childhood are reflected in a mystical experience that took place in 1918, four years before the Vienna episode (Prevelakis, 1965). As Kazantzakis (1961) describes it in his *Report to Greco*, one night, while walking with a childhood friend on a snow-covered hill in Switzerland, he was seized by a sense of drunkenness that made the landscape around appear unnatural, and at the same time very familiar—a *déjà vu* experience. Suddenly, as the trail turned and revealed the dim lights of a distant village, ". . . an astonishing thing happened to me. I still shudder when I recall it. Halting, I shook my clenched fist at the village and shouted in a furor, 'I shall slaughter you all!' A raucous voice not my own! My entire body began to tremble with fright as soon as I heard this voice" (p. 25).

Kazantzakis attributes the experience to the activation of the violent nature of his collective unconscious, referring to his ancestors on father's side as "bloodthirsty pirates" and "warrior chieftains." In describing his conflict around this timeless violence, he writes:

> Now and then, sporadically, a sweet voice sounds in the very center of my heart: "Have no fears. I shall make laws and establish order. I am God. Have faith." But all at once comes a heavy growl from my loins, and the sweet voice is silenced: "Stop your boasting! I shall undo your laws, ruin your order, and obliterate you. I am chaos!" [p. 26].

Kazantzakis and the Oedipal Mother

As suggested in Chapter XIII, some mystical experiences may be the result of a sudden reactivation of aggressive wishes against the father of early childhood. While in Vienna, Kazantzakis was more concerned with the violence of his feelings toward the oedipal father than about his spiritual solitude or his unfulfilled sexual urges. What frightened him and made him feel helpless was the unconscious equivalent of a libidinal attachment to his mother, reactivated by the chance encounter with a seductive woman in the darkness of a foreign place, and the associated anger at his father, reinforced by a sense of imminent violence that, as with the environment of his early childhood in Crete, pervaded the atmosphere of post-World War I Vienna. Sexually inhibited toward his wife, but deeply impressed with her common sense, competence, and courage, he pleaded with her to come and share his self-imposed exile in order to protect him from the re-emerging violence of oedipal object relations as a mother substitute, which his wife, Galatea (and later Helen), apparently was for him. As Kazantzakis (1961) confides after finishing the portrait of his father in the *Report to Greco:* "My mother was a saintly woman. How was she able to feel the lion's heavy inhalations and suspirations at her side for fifty years without suffering a broken heart? She had the patience, endurance, and sweetness of the earth" (p. 34).

But, so far as he remembered, she never laughed—except once. It was Spring and the family had gathered in a large room for a festive occasion. A torrential rain broke out. As the poet, a boy of three or four, recalls:

> Outside the house God was still roaring. The thunder increased, the village's narrow lanes had been transformed into rivers, the stones tumbled down them, laughing wildly. God had become a torrent; He was embracing, watering, fruitifying the earth.
>
> My father turned to my mother. It was the first time I ever saw him look at her with tenderness, the first time I heard sweetness in his voice.
>
> "Sing Marghi," he said to her.
>
> He was giving her permission to sing, giving it in front of all the other men. I became angry, though I don't know why. Rising all in a ferment, I started to run to my mother as though wishing to protect her, but my father touched my shoulder with his finger and made me sit down. My mother seemed unrecognizable; her face gleamed as if all the rain and lightning were embracing it. She threw back her head.

> I remember that her long raven-black hair suddenly became undone and fell over her shoulders, reaching to her hips. She began. . . . What a voice that was: deep, sweet, with a shade of throatiness, full of passion. Turning her half-closed eyes upon my father, she sang a mantinadha which I shall never forget. I did not understand at the time why she uttered it, or for whom. Later, when I grew up, I understood. Looking at my father, her sweet voice filled with restrained passion, she sang: "I'm amazed the streets don't blossom when you stroll, And you don't become an eagle with wings of gold." I looked the other way to avoid seeing my father, to avoid seeing my mother. Going to the window, I pressed my forehead against the pane and watched the rain fall and eat away the soil [p. 39].

The Wish to Save God

As a small child, Kazantzakis prayed to God to make him God. As a youth, he studied the teachings of Tolstoy, hoping one day to succeed where the Russian novelist had failed: "to create religion" (Prevelakis, 1965, p. 7). Shortly after the Vienna crisis, he decided to teach himself Russian and the art of carpentry in order to go to the Soviet Union and "try the Word that I carry within me" (Kazantzakis, 1958, p. 120), a blend of Tolystoyan and Bergsonian[5] ideas with which he sought to counter Nietzsche's blasphemous assertion that God was dead. Kazantzakis declared that his mission was "to save God," which can be easily interpreted as deriving from the timeless wish to kill the awesome father of his childhood, whom he continued to hate and fear throughout his life.

In a confidential interview a few months before his death, Kazantzakis spoke about the close affinity he had once felt with the person and mission of Christ:

> Twenty years old [he recalled], I had written a tragic play, "Christ," then a poem, "Christ," but Christ did not

[5]In earlier years, Kazantzakis attended Bergson's classes at the Sorbonne, and translated some of the philosopher's work into Greek, counting him along with Christ, Buddha, and Lenin among the heroes of his life.

> cease to dominate every intellectual concern of mine. It was like a cyst that they take out and grows up again. Then I got the idea of a novel, which has not been written yet. . . . A young lad came to Nazareth, ideologist, full of enthusiasm, vegetarian, pacifist. He didn't touch women and wanted to save the world. There are so many who are not well balanced, who are neurotic, as they call them today! To make the story short, he no more and no less wanted to become the Messiah. His poor mother wanted him to get married, but the idea made him mad. Finally, they persuaded him to see a rabbi therapist and psychoanalyst, and there start the usual questions. You know those psychoanalysts! I went to see one of them once in Vienna. . . . Well, let's go back to our young lad. The rabbi cured him, that is he made him give up his ideas, and he became the best cabinet maker of the village. He also had a specialty, he could make wonderful crosses, upon which they crucified those who raised head against the Romans [Jouvenel, 1959, p. 239].

Kazantzakis died before he could write about the young ideologist who was cured by the Jewish psychoanalyst. But he returned to the subject of Christ in two of his last novels: *The Greek Passion* (1948), in which a young villager, Manolios, while playing the role of the Messiah is seized by the desire to sleep with the prostitute of the village—herself playing the role of Mary Magdalene—but is prevented (saved, as Stekel allegedly put it), by the sudden appearance of the same skin disease that crippled Kazantzakis in Vienna; and *The Last Temptation of Christ* (1955), in which the crucified Messiah dreams he is living his life over again as a common mortal and, freed from his neurotic scruples, manages to sleep with Mary Magdalene.

Kazantzakis' latent hostility against his father, rekindled by the regressive wish to sleep with his mother, brought back to the surface the fear of the oedipal father and the threat of castration, which the swelling of his face not only symbolized, but essentially materialized with its crippling effects.

The Influence of the Environment

At this point, I might mention another factor that probably contributed to the awakening of Kazantzakis' aggressive instincts and the fear that rises from the awareness of one's own potential for violence: the influence of the crowd, threatening and at the same time seductive, with which he found himself in close contact a little before the onset of his symptoms. Here is how the writer himself described the circumstances that, creating a state of psychic automatism, led him to the fatal encounter with the woman of his forbidden childhood fantasies:

> Going outside to the street, I mingled with a wave of people all running somewhere in a great hurry. I ran with them; I no longer had anything to fear. They are wraiths, I reflected, a mist composed of dewdrops. Why be afraid of them? Why not go along and see what they're doing? Reaching a movie theater with red, blue, and green lights, we went inside and enthroned ourselves in velvet-cushioned seats. At the far end was a bright screen over which anxious shadows were hurriedly passing. What were they doing? Kissing, killing, being killed. Next to me sat a girl. . . . [Kazantzakis, 1961, p. 352].

A few hours later, as he is lying in bed alone, full of anxiety about the strange swelling of his face, Kazantzakis' mind goes back to the island of his childhood where leprosy is endemic, recalling the repulsive sight of the hapless victims and his own secret wish to have them all killed. And now he feels that God has struck him with the terrible disease, punishing him for his insolent childhood wish.

The Mystical Solution

Kazantzakis' violent id is reflected in his literary work, particularly in *The Odyssey* (1938), a litany of orgies, mass slayings, pillagings, rapes, and other fantastic, often supernatural, adventures; but violence hardly ever appeared in his behavior. Most of those who knew him describe him as exceptionally kind and calm. Only his first wife, Galatea, and to a certain extent her brother, Lefteris (Alexiou, 1956), remember him as cruel, egocentric, and cowardly. Interestingly, in his letters to Galatea, Kazantzakis (1958) describes himself in the same critical way.

Thus, after complaining about his loneliness and his "psychosomatic" illness ("that psychic intervention of mind in the body" [p. 55]) which made him look like an abominable leper, he writes: "It seems to me that with all the cruelties, meanness (of mine), etc. to which you in particular (perhaps only you) have been subjected, there exists inside me something very kind and meek" (p. 56).

That he so often talks about hatred almost for no other reason than to denounce it should be significant. In spite of the fact that he talks constantly about love—his love for Galatea—and about the wish to become, like Buddha, full of acceptance for man's cruelty and meanness, it is not hard to understand how in his loneliness, abandoned as he felt by his wife and some of his closest friends,[6] hardly able to speak the language of the Viennese masses whose suffering and anger he identified with, Kazantzakis was seized by a great fear. It was the fear of his own violent nature, which threatened to erupt as he could not transform it into creative, social, or artistic activity, until he decided to follow Stekel's advice and leave Vienna. As he himself put it in closing the chapter of his illness in the *Report to Greco:* "Ever since that day I have realized that man's soul is a terrible and dangerous coil spring. Without knowing it, we all carry a great explosive force wrapped in our flesh and lard" (Kazantzakis, 1968, p. 357).

His early identification with Christ helped Kazantzakis to accept his sexless existence as appropriate; to see sex, if not literally as sinful, as a temptation away from his life-long ambition to create a new religion of spiritual love and universal peace. At the time he arrived in Vienna, he was trying to replace in his consciousness Christ with Buddha, a Godlike figure characteristically more concerned with the renunciation of war and violence than with love.[7] In my view, Kazantzakis' effort to identify with Buddha's image and ascetic principles, rather than being the cause of his anxiety—as he seemed to believe—was a way of coping with the anxiety caused by his inner violence and the

[6]Especially his former companion, composer Sfakianakis.

[7]In this respect, compare Bertrand Russell's identification with Buddha in Chapter X.

threat to erupt in the midst of the seductive, outwardly peaceful but inherently cruel and hostile environment of Vienna. Whereas he had been successful on previous similar but less exciting occasions, now he failed. Asceticism and the attempt to gain a mystical experience by composing an epic poem about Buddha, which Kazantzakis' (1927) himself attributed to an inner struggle against a "tremendous" self-destructive force, were of no avail. Kazantzakis' psychic balance was saved by means of a psychosomatic symptom which, as Stekel correctly observed, has been a common defense among ascetics of all times and places. A more vulnerable ego than Kazantzakis' might have given way to psychosis.

Kazanzakis tried to resolve the conflict around aggression and the fear of his own violence by assigning to his violence and that of the world an instrumental role—that of serving universal love and the attainment of a noble consciousness, which he identified with the salvation of God. Unlike Nietzsche, who spoke of evil as the supreme goodness, Kazantzakis identified it with the world of the prevailing social system—capitalism—denouncing it angrily and, eventually, in the model of the mystical tradition, denying it along with the reality of goodness, seeing both evil and good as two illusory aspects of one and the same reality, experienced as selfless, timeless, and spaceless existence, or God.[8]

[8]Kazantzakis (1927) outlined his mystical view of the world in his *Asketike* (translated into English with his approval as *The Saviors of God: Spiritual Exercises*), which he started composing shortly after his crisis in Vienna but kept revising, like *The Odyssey,* which embodies the same ideas, for many years. Its final version, published shortly after the end of World War II in Greece, closes with an incantation of nine messages in capital letters, the last three of which clearly echo Nietzsche's Zarathustra.

XIV
TIME AND DEATH

THE CONCEPTS OF TIME AND DEATH are inseparable, the one has no meaning without the other. As Eissler (1955) put it: "Death and time are indelibly connected in our subjective experience. If man could not experience time, he could never fathom death, which would then be as meaningless to him as it is to the animal" (p. 265).

Time objectifies human experience, keeps score of a person's life, delineating its limits and its stations, its progress, its existence. The first thing surgical patients awakening from anesthesia want to know is the time of day: "What time is it?" And as they forget the answer instantly, they keep asking until full consciousness is restored. As Dooley (1941) points out, "To know the time is to reestablish contact with the world of the living" (p. 18).

The Fear of Death

Paraphrasing Falstaff, Freud (1915) liked to say: "everyone owes nature a death and must expect to pay the price" (p. 289). But unless one is faced with a life-threatening situation or is suffering from a neurotic illness, one thinks of death as something out there, as a dark mystery that lies in the indefinite future, as a cold reality whose time and place are determined by some strange, heartless and deaf deity—Charon, old, skeletal man with a scythe, or inexorable Atropos with her spinning wheel—whose fatal gesture no one can prevent or delay.

> A middle-aged man, resuming psychoanalysis after a period of hospitilization and home rest for a heart attack, said: "I am angry that such a thing should happen to me. I had no such

> Conceptualizing the future when I know that it is going to be cut off by death is not easy." To his analyst's comment that everybody's future was going to be cut off by death, the patient responded sarcastically: "Yes, and I do not have to have a heart attack in order to know it, if that's what you mean. All men are mortal, Socrates is a man, therefore Socrates is mortal. But mortality is as much of an abstraction as is immortality, and I happen not to believe in life after death. And I'm not Socrates not to mind dying. Besides, he was an old man; I wouldn't care that much if I were his age. You imply that I was denying things before, the idea of death, the fact that death is in all of us, sooner or later he takes over. But why sooner, why me?"
>
> When the analyst asked him to associate to the idea that it was unfair to be the one to have a heart attack, the patient came to identify his anger with the experience of being punished by his father while his younger brother usually got away, "scot free." He recalled how enraged he was at his father, especially at the idea that he could do nothing about it, hit back or get away, because of the fear of greater punishment. He felt his father was unfair and insensitive, as he had experienced his analyst a few moments ago; but at least he could talk back to his analyst, something he could never do with his father. He also recalled feeling angry at his father for ridiculing him when he felt afraid of the dark. At this point, the analyst wondered whether the patient's father ever punished him by confining him in the basement or other dark place. Without excluding the possibility, the patient denied having such a memory. But being literarily minded, he volunteered the information that Ingmar Bergman, the Swedish film director, was punished as a child by his father that way, being locked into a dark closet, at which point the patient wondered whether Bergman's early relationship with his father might explain why his films were permeated with the sense of morbidity and death.

Philosophers have argued that the fear of death is irrational. As Democritus put it, why should you be afraid of death? When you exist, death does not; and when death does, you do not. Freud (1915) said: "It is indeed impossible to imagine our own death; and whenever we attempt to do so we can perceive that we are in fact still present as spectators" (p. 289).

According to Heidegger (1927), the idea of death is really dreadful only when one tries to grasp the notion of one's own nonexistence, the idea of nonbeing. In contrast to the ordinary notion of death, which is based on the conception of oneself as a dead object—very much in the sense that one thinks of the death of a love object—the awareness of "not being" elicits what existentialists, beginning with Kirkegaard (1844), have described as the sense of dread, the fear of annihilation. I imagine that is similar to the feeling of dread experienced during a schizophrenic breakdown, a feeling that a psychoanalyst, limited by the notion that everything is reducible to a previous experience, may not be able to understand or interpret.

> I remember once waking up from a dream by an extraordinary experience, what I can best describe as a soundless, colorless, painless explosion, a blank screen of terror. I remember nothing from the dream that preceded the experience, except for the image of an older friend from Germany, who had been in Hitler's occupation army in Greece and who, in retrospect, might have stood for the oedipal father of my childhood. But I felt that the terror of the experience that followed the abrupt disappearance of the dream had nothing to do with it, except for shattering it like a picture window behind which there lies only a dark, empty space, blowing it out of my vision while I was still asleep with eyes closed, not even half awake. I wonder, could that be the prototype of the experience of death, or at least of the experience of dread that Heidegger describes in connection with the idea of one's personal death, the conceptualization of one's own absence, the awareness of not being?

According to Klein (1932), the prototype of anxiety is the fear of one's own innate aggression, the awareness of Freud's death instinct, the potential destructiveness that we all carry within us and can be directed against someone we fear and hate —the oedipal father, or our own helpless self. On the other hand, there is the fear of death as a phenomenon, the confrontation with the lifeless body of a person or animal unaware of one's presence, indifferent and unresponsive—"the fascination of the corpse... the worst purveyor of anxiety" (Malraux, 1974, p. 139).

Freud may have been preoccupied with the idea of death, predicting his own end prematurely several times (Schur, 1972), yet he reasoned that the fear of death was not what it appeared to be. As he told Marie Bonaparte (1940), how could one be afraid of something one has never experienced? According to Freud (1923):

> . . . death is an abstract concept with a negative content for which no unconscious correlative can be found. It would seem that the mechanism of the fear of death can only be that the ego relinquishes its narcissistic libidinal cathexis in a very large measure—that is, that it gives up itself, just as it gives up some *external* object in other cases in which it feels anxiety. I believe that the fear of death is something that occurs between the ego and the superego. . . like the fear of conscience, [it is] a development of the fear of castration [p. 58].

Freud (1917) also pointed out that in severe obsessional neurosis and in melancholia the fear of death becomes overwhelming because of the intensification of guilt that follows the loss of an ambivalently held love object and the introjection into one's superego of the hatred that was formerly identified with the love object.

Consistent with the role that man's conscience—our superego—plays in the conceptualization of time and death is Dodds' (1951) observation that the psychology of Homeric Greeks, Western man's cultural ancestors, was characterized principally by shame rather than guilt—guilt and the concept of conscience appearing for the first time in the plays of Sophocles and Euripides in classical Athens. Homeric Greeks, at least as depicted in the *Iliad,* had virtually no interest in temporality, absolute or relative. Thus, as de Romilly (1968) asserts, time is never the subject of a verb in Homer, and the idea of time does not appear to have been important before the advent of Greek tragedy. It was only with Euripides that the destructive aspects of time were acknowledged, and with Sophocles that time's power of alteration was lamented. And, for the first time in literature, we hear expressions permeated by the consciousness of time as duration: "the time has come"—"now is the time"—"this is the decisive time"—"will she know in time"—"just in time."

The Greek's relative immunity to feelings of guilt, even in classical times, and their present-oriented attitude seem to go along with an unconcern for violence and death, which remained as prevalent as in Homer's prehistorical world. In contrast to their great respect for freedom and civil rights, Greeks on the whole displayed an uncanny indifference to the idea of violent death and the value of human life, as it is made strikingly clear by Thucidides in his history of the Peloponnesian war.

The Wish to Kill Time

According to Rank (1914), one of Freud's earliest and most intuitive students: "the fear of becoming old. . . is really the fear of death" (p. 77). The fear of old age may not be tantamount to the fear of death, as Rank and more recently Eissler (1955) contended, but the wish to stay perpetually young and to postpone old age indefinitely is basically the wish to postpone the inevitable: death. Whatever else goes with old age—illness, misery, loneliness—like virtually anything in life, can be tolerated. "Men get used to anything, except dying," said Malraux (1974, p. 47). Man's uncomfortable awareness of the passage of time and his wish to make it go faster, stop it, somehow reverse it, or eliminate it from consciousness and with it the awareness of change, betrays the unconscious dread of failure, separation, and death. This ubiquitous triad of evils prompts the regressive wish to "kill time," to go back in time, to search for time past, for memories of a presumably happier youth and childhood, the golden dream of mother's breast, the oceanic experience of union with mother, the luminous, ineffable, timeless experience of mystics and drug addicts throughout the ages and the world.

"Killing time" is a less incestuous and more narcissistic enterprise than the wish to kill one's own childhood father and to claim mother for oneself on the basis of preoedipal, symbiotic memories or fantasies. The desire and pursuit of recreation and adventure, commonly thought of as relieving the tedium of everyday routine and toil, may be viewed as the compulsive manifestations of the wish to escape the unconscious awareness of a long series of evils time has perpetrated upon us: oral and anal and oedipal frustrations or traumata, experience of separation and loss during childhood; frightening memories of violence and seduction, beginning with that of the primal scene; more or less repressed murderous or incestuous wishes and castration threats involving parents and siblings; real or imaginary evils that accumulate in time and emerge occasionally in a hot summer's nightmare or are excitedly glimpsed in a bedtime novel, in a time and space other than our own. Whether watching a second-rate movie or an artful theater performance, visiting a museum or an amusement park, swimming in the blue waters of the Mediterranean or going on an African safari—time is manipulated, annihilated in its forward progression, recaptured, relieved, undone, killed.

The Wish for Eternity

The fear of death and its derivative belief in the immortality of the soul are inherently related to the concept of time. As the awareness of time is presumably peculiar to man, so is the idea of personal death. And as man knows he must die, he is afraid of time as he is of death, contemplating it with anxiety or counterphobically denying it.

Pointing out that attempts to deny the fear of death are universal, as the belief in the immortality of the soul might indicate, Sterba (1968) suggested: "The mystic state with its hallucinatory, wish-fulfilling revelations provides the strongest opportunity for grandiose denial of our perishability" (p. 84).

Chemical substances such as LSD, marijuana, and alcohol, as well as naturally occurring alternate states of consciousness, including dreaming and religious conversion, have the power to distort the sense of time in ways that are characteristic of these substances or states of being. The same is true for psychiatric illness. Consciously or not, man is inclined to change the rhythm of his temporal experience or to altogether escape from it—and that may be a function of his wish to escape the awareness of violence, mainly his own, and the threat of helplessness and death, what Schiffer (1978) has broadly labeled "the trauma of time." The need for the arts, poetry and music in particular, with their appeal to the mystical, nonrational part of the mind, may to a large extent be a manifestation of the universal need to escape from the boundaries of time, to exchange one's sense of time with that of timelessness, to transcend from temporal existence into eternity. Such is religion's main attraction, the promise of a land where, as the Angel in St. John's revelation puts it, "There shall be time no longer."

But eternity is of two kinds: timeless and timeful. Timeless is the eternity of euphoria or oblivion, the eternity of blissful, ecstatic, mystical experience, the state of consciousness that prevails when one is in love and in firm possession of one's love object, a condition of intoxication that may be also brought about artificially by means of alcohol or drugs. As Loewald (1978) put it:

> Philosophers and theologians have spoken of the *nunc stans,* the abiding now, the instant that knows no temporal articulation, where distinctions between now, earlier, and later have fallen away or have not arisen. All of us know, I believe, poignant moments that have this timeless quality: unique and matchless, complete in themselves and somehow containing all there is in experience. As experience augments and grows in an individual's life course, these instants, *in* time but not *of* time, contain more and more meaning which is poured into the *nunc stans* in such a way that temporal and other articulating differentiations are dissolved or become condensed into oneness. What was lived through earlier and later, and the mental categories of secondary-process mentation—all fall away, collapsing into an instant, into that one experience which then stands for all experience, although only "for one instant" [p. 65].

On the other hand, there is the eternity of pain or anguish, the eternal present, when neither future nor past make any sense, in which both future and past are meaningless, unable to relieve the pain and the anguish of the present that feels like eternity. As Emily Dickinson (1890) wrote:

> Pain—has an Element of Blank—
> It cannot recollect

When it begun—or if there were
A time when it was not—
It has no Future—but itself—
Its Infinite contain
Its Past—enlightened to perceive
New Periods—of Pain [p. 323].

Pain condenses eternity into a single-minded present, time coming suddenly to a standstill:

Pain—expands the Time—
. . . . . .
Pain contracts—the Time—
Occupied with Shot
Gamuts of Eternities
Are as they were not— (p. 452).

Pain is a human reality measured in time—slow or arrested time—the only reality that lasts. Slow or arrested time is the only real time. Fast time is in one's consciousness selectively, off and on; it exists periodically or in retrospect, as one stops and thinks about it; it is mainly felt as the prevailing wish, need, or desire. When the fear of death becomes intense, time defensively slows down or stops.

> In the acute terror of death the ego experiences itself as having reached the point when time comes to a standstill and the whole future has been reduced to the pinpoint of the last moment of life. The last moment is visualized as being deprived of duration (biological present) [Eissler, 1955, pp. 277–278].

St. Augustine's elaboration of the Platonic definition of eternity as standing and not passing and having in itself no past or future would sound more like the eternity of pain or anxiety, if it were not for the fact that there is no reference to the present in it. It is my feeling that Plato would not conceptualize eternity in terms of negative movement or space (as standing and not passing), when he meant it to be one of those ideal concepts that transcend man's temporal existence and pain. As Eissler (1955) suggested, St. Augustine's version of eternity is more likely to be based on the model of time as experienced during orgasm. Overwhelmed by sensuality, the ego fuses with the body and the world "into one complex, indivisible experience inaccessible to further analysis [which] makes it possible for time to become pure present" (p. 271). Which may also explain why some neurotic individuals are afraid of having an orgasm, as others are afraid of going to sleep—because they are afraid of dying.

The association of death and dying with the idea of sexual orgasm is not so uncommon, as the incidence of accidental suicide by hanging may attest. Contemplating the imminent possibility of his own end from a mysterious neurological illness, the aging Malraux (1974) recalls the words of an Indian friend:

> Westerners, and they are not alone in this, believe that there exists one room, which is life, and another, which is the beyond, and that death is the door through which we pass from one to the other. Isn't this so? Why do they dramatize the door? Death is the path to Enlightenment. You know this when you come back from it—or at least from something like it. My guru used to go into a trance for several hours, and come back. The Buddha experienced Enlightenment well before dissolving into it. One can choose one's death [pp. 86–87].

The symbolism of death as a narrow door ushering into a new, ecstatic world beyond is suggestive of sexual intercourse and its orgasmic experience. The Victorian view of death as a pleasurable release and as a supreme accomplishment cannot be explained on the prevalent belief in the immortality of the soul alone. The following exchange between William James and his sister Alice shortly before she died from breast cancer at the age of forty three is, I think, illustrative of the connection. James wrote:

> When that which is *you* passes out of the body, I am sure that there will be an explosion of liberated force and life, till then eclipsed and kept down. I can hardly imagine *your* transition without a great oscillation of both "worlds," as they regain their new equilibrium after the change! [Yeazell, 1981, p. 186].

Alice replied that dying was "the most supremely interesting moment in life" (p. 186).

The idea of one's own death, its proximity or even certainty, is not enough to make one afraid of death. People go to war, face the execution squad, or commit suicide without fear. Socrates chose to drink the hemlock, trying to cheer up his friends and disciples in the process rather than allowing them to lead him to what he felt was going to be an undignified existence as a fugitive from the law of his native land. Such people do not necessarily believe in immortality (Socrates had his doubts),[1] but I imagine time for them becomes meaningless; they are able to deny it or to transcend the sense of time on the occasion, like people in love or under the influence of certain drugs. Intense emotional investment in a person or idea—love, conviction, a cause—has the power to cancel the consciousness of time and hence the fear of death. As Wittgenstein (1921) whimsically observed: "If we take eternity to mean not infinite temporal duration but timelessness, then eternal life belongs to those who live in the present" (p. 147).

Hence the glorification of the here-and-now (or Now) by mystics of all persuasions, and the belief in an eternal God, in a Being that was present before the beginning of Time, before and after Death.

[1]According to Bertrand Russell (1945): "[Socrates'] courage in the face of death would have been more remarkable if he had not believed that he was going to enjoy eternal bliss in the company of the gods" (p. 142). But, I suspect, Russell was just jealous.

XV
TIME AND THE LIFE CYCLE

"Every experience," wrote Straus (1947), "receives its specific significance, its special value, from its temporal position. Even in its content each moment is determined by its place in our life history. We all know that we will have to die" (p. 256). But it is only gradually, and more or less imperceptibly, that our consciousness becomes permeated by a sense of personal destiny—a destiny that leads inexorably to decay and death.

A Child's Time

Experientially as well as clinically time is a metaphor for one's state of well-being in a world of change that deals with other human beings and reflects the way one feels about oneself and others—more specifically, the quality of one's internal object relations and ego integrity or strength. If time, in a descriptive sense, is "a coordination of movements or speeds" as Piaget (1970, p. 59) says, or, as I think is more accurate to say, the perception of contrast in motion (as space is the perception of contrast in position), in a psychodynamic sense, time is the perception of contrast between the self-image and object images, the perception of a discrepancy generated by object-directed drives creating intrapsychic tension. As it rises or falls, this state of need-tension is experienced as an inner sense of movement or speed that, projected on the outside world of objects, becomes the experience of time as duration.

A child's concept of time as perspective, on the other hand, and in particular the notion of future, develops around the capacity to delay gratification—the ego's increased adherence to the reality principle. The younger the child, the shorter the delay; which makes the future seem to be further away. A child's widening temporal perspective, stimulated by his identification with older siblings and playmates, remains limited to images of later childhood and adolescence. His conceptual grasp of time as a measure of change and of the future as a realm of possibilities can be traced in the maturation of the ego apparatus through the various stages of psychosexual development—oral, anal, genital—with its epigenetic, sociocultural challenges and conflicts.

The emergence of the superego as an autonomous system of the mental apparatus at the end of the oedipal period contributes significantly to the child's adaptive conceptualization of time, influencing his or her ability to endure stressful experiences, to schedule decisions, and to wait; to be punctual, and to synchronize his or her activities with those of peers and grownups. Characteristically, psychopathic behavior in children is correlated with a defective conceptualization of time (Bender, 1950). Delinquent youngsters do not remember the past, and consequently cannot benefit from their past mistakes; they do not concern themselves with goals and the future, and therefore cannot be motivated to control their behavior for future advantages. The concepts of distant future and adulthood, however, remain vague, unrealistic, and in essence disconnected from the present reality for all children, normal or not, until the period of latency is over.

Time and Adolescence

A broader perspective on the future is reached in adolescence, with its emergent sense of self- or ego-identity. A heightened experience of affects, brought about by endogenous, hormonal factors and a rapid change in body image, forces upon the adolescent the awareness of time perspective as a personal, creative, making or breaking destiny. It is the task of the adolescent to integrate his or her self in the three dimensions of time, to create a personal perspective out of what appears like an overwhelming succession of present wishes and images, to see that the past does not become a false memory all but drowned in the urgent wishes of the present, and to see that the future does not become an empty promise suspended above and beyond the vivid imagination of the present. It is the adolescent's task to channel his or her rekindled oedipal wishes and fears outside of the family context, away from the incestuous immediacy of his or her home environment, sublimating them into romantic relationships and peer activities of a competitive or cooperative nature, into pursuits geared to social expectations and future fulfillment, and if successful to experience a time bound identity.

Through his newly acquired ability for "formal operations" (Piaget, 1936), the young adolescent can now deal logically with the concept of the distant future, with images and possibilities expressed in abstract terms as hypothetical and far removed from the here-and-now reality. More specifically, the advent of adolescence makes it possible to integrate distant putative events with past and present experiences, and for ego identity to acquire stability and coherence within the continuum of time. For the first time the adolescent person grasps the idea of life as a temporal unit, a narrative with a beginning and an end.

The ability to perceive the future as predictable or open with meaningful possibilities, specifically the possibility of choice, depends on the ability to perceive the present as the result of a consistent, generally self-fulfilled past. His interactions with the adult world create conflicts and frustrations that, like the child, the adolescent tries to solve by resorting to actual compensations. But, in addition, he plans future compensations in terms of social reforms or scientific discoveries (Inhelder and Piaget, 1955). On the other hand, adult pressures, aimed at shaping his thinking in terms of social roles that are geared to the future, may lead a young person to the espousal of an anti-future, anti-establishment attitude that glorifies action and the pursuit of immediate satisfaction. There is often "a sense of great urgency and yet also of a loss of consideration for time as a dimension of living" (Erikson, 1956, p. 82). Even when renouncing it, however, the adolescent becomes imbued with a sense of responsibility, the awareness of the present's role in terms of the future. The anticipation of the unknown and the anxiety that such anticipation entails, in particular as it concerns the mystery of death and the realization that one's life must come to an end, shape the present image of the future as well as that of the past.

Time and Adulthood

"Generativity vs. stagnation," to use Erikson's (1950) felicitous terms, define the normal adult's ethos — the sense of accomplishment and purpose that comes with the development of a career and a family life of one's own. So long as one is immersed in the pursuit of one's goals, one acts in the present according to what one anticipates to happen in a relatively distant future. But, as Cottle and Klineberg (1974) point out, "unless images of the future give rise to present emotions, anticipations will have little effect on a person's behavior" (p. 16). The motivating power of the future may be described as a process of self-projection whereby the individual is able to transpose the image of himself into a probable situation, experiencing in the present something similar to what he might experience if he ever found himself in such a hypothetical situation. This ability may be defined as empathy, in which case the object with whom one establishes an empathic relationship is one's own self in an imaginary future situation. Such an ability can hardly exist if one is not able to empathize with another person. Kernberg's (1976) observation that narcissistic people "have difficulty evoking not only real people in their past but their own self experiences with such people" (p. 73), being generally applicable with regard to the future, may explain why severely narcissistic individuals are unable to maintain a consistent and meaningful sense of time.

Normally, one's adult time perspective should be determined by one's personal identity and commitment to a social role. A stable sense of identity, reflecting a relatively nonconflictual internal object world and a comfortable, reassuring and reliable self-esteem, allows a person to perceive the present as an integral part of a continuum that, like a free-flowing narrative or a well orchestrated score of music, enfolds effortlessly and predictably, without devastating regrets or shocking surprises. In such a continuum the past is recalled or recognized effortlessly with an appropriate sense of guilt, and the future is anticipated predictably with an appropriate sense of anxiety. With such a continuum as the background, one experiences the flow of time on the whole evenly, as a stream of consciousness in which one's self keeps moving along with time, alert and engaged in people, ideas, activities whose meaning is grounded on spontaneous interpretations of one's personal past or being, and is constantly restructured or revalidated by new experiences that define a personal future or becoming.[1] Within such an ideal frame of reference, deviations in the way time is experienced depend on individual differences in ego structure and character style, and of course on special life events and environmental circumstances.

Falling in love, getting married, having a child, losing a love object, studying for an examination, winning a promotion, facing an emergency are ubiquitous yet special situations that alter one's experience of time, temporarily, predictably, yet impressively. A person's particular ego apparatus, instinctual drives, defensive organization, and superego play a role in the direction of one or another extreme represented in pathological conditions, neurotic, psychotic or characterological, specifically and more than temporarily. The following example illustrates how a basically normal individual may experience time in a relatively distorted way when confronted with special life events:

> A thirty-year-old schoolteacher, in psychoanalysis for four years and while contemplating the possibility of termination, began worrying about getting old. He felt he was behaving like a middle-aged man, having the fantasies of a middle-aged man, tempted by teenage girls, his own students. He wondered whether he was ever going to be really mature, dignified and secure, as he felt his analyst was and wanted him to be.

[1]As Allport (1955) points out: "A great many states of mind are adequately described only in terms of their futurity. Along with *striving*, we may mention *interest, tendency, disposition, expectation, planning, problem solving,* and *intention*. While not all future-directedness is phenomenally appropriate, it all requires a type of psychology that transcends the prevalent tendency to explain mental stages exclusively in terms of past occurrences. People, it seems, are busy leading their lives into the future, whereas psychology, for the most part, is busy tracing them into the past" (p. 51).

> I suggested that he was perhaps not so much averse to the idea of growing old but looking forward to it as a solution to his feeling guilty about unacceptable sexual and, for that matter, sadistic fantasies. He protested. Meaningful as the analyst's interpretation might sound, it did not correspond to the way he felt about the passage of time, for, along with the idea of growing old, he had the feeling that time was "going out fast." He used to feel he had all the time in the world, that he was never going to grow older, that he would always be the same. But now things changed too fast, especially during the last year since his first child was born. His child had gone through so many changes so fast in front of his eyes that he could no longer feel the same. He was thinking, there are 52 weeks in a year and 520 weeks in ten years, 3,650 days, and whereas before that would have sounded like a long time, now it was too short. He knew he was growing older, which would be fine if he could feel more responsible, more dignified, more secure. He wanted his analyst to reassure him that he could be all these things before he terminated his analysis, an idea about which he admitted feeling very anxious.

Normally, time seems to pass slowly when one anticipates a greatly desired event; and, vice versa, time advances rapidly when one faces a future for which one feels unprepared. Again, time seems to pass rapidly if it is filled with interesting events or need-fulfilling, happy activities; time seems to pass slowly if filled with dull or frustrating experiences. Yet, in retrospect, one tends to feel the opposite—the experience becomes paradoxically reversed. A rather short time spent in a lively fashion is remembered as long, while time that has been experienced as long and monotonous is remembered as short. The explanation for this apparent discrepancy may lie in the fact that, when time seems to pass rapidly, one's attention focuses on the events that take place during that time and hardly on the passage of time until time is over, and vice versa: one is likely to remember as longer the period whose content held one's attention most. Such is indeed the classical view (James, 1890; Fraisse, 1963), according to which the length of apparent duration depends on the number of changes or individual elements experienced in a particular interval of objective, clock or calendar time. The more eventful it is, the longer a time period appears to be; the more one is able to recall from its content, the longer one's estimate of its duration will be.

When we lose interest in things, when our emotional ties with the world become loose or severed, we feel that time is inordinately long and heavy, that time has slowed down or stopped. It is, to borrow Bergson's (1889) metaphor in describing a consciousness empty of sensations and ideas, like hearing foreigners talking to one another in a language we do not understand—they seem to speak very loudly "because their words no longer call up any ideas in our mind, and thus break in upon a kind of intellectual silence and monopolize our attention like the ticking of a watch at night" (pp. 40–41). Time becomes loud like meaningless words, words spoken by people who mean little to us emotionally. Characteristically, we find it very hard to project ourselves into the future. Feeling embedded in a present that is devoid of interest, we perceive the future as a projection of the present and therefore as no future at all.

On the other hand, when we are emotionally involved, whether in action or in people, we experience time as going fast and ourselves as going along with it. This is particularly true when we are in love.

> A professional woman, married and mother of three children, in psychoanalysis for about a year, while under the impact of an intense erotic transference, said: "My God, I am already forty, I'll soon be eighty or ninety, time is running out." And, a little later: "What day is today? Thursday? Time is passing so quickly this year, especially this spring."
>
> "I wonder what makes you feel that time passes so quickly?" said the analyst.
>
> "It is so!" the patient continued. "I'm looking back and it seems as if the days pass in no time at all. It's because I'm coming here. Last year time didn't pass at all. It was really awful. I feel really funny, just tired . . . strange." (Silence) "I'm getting very frustrated. I saw your wife at the parking lot and I don't know why, but I'm very upset about it. I think that's why time goes so fast. That's why I wear a different dress every day—for you. Time goes fast when I come here, and time doesn't when I don't.

Like the one week away from you, time would not move. I was saying, why do I want time to pass so quickly when I'm so concerned about getting old."

Time and Middle Age

> When summer comes, I go back to Greece, my birthplace and childhood playground. It used to be that landing at the airport in Athens made my heart pound with anticipation: "One more summer in Greece," I would say. Then one summer, as my plane was again landing, I sensed in me a certain sadness. I looked back and said: "One less summer in Greece."

After a certain age, the realization that our span of life becomes shorter enhances the feeling that times goes faster. As we compare our age to normal life expectancy, our time perspective narrows down.

A vivid description of the changing experience of time as one enters middle age was given by Virginia Woolf (1976) for whom, as with other creative but neurotic people, growing older was a painful reminder of an unfulfilled self:

> Every ten years brings, I suppose, one of those private orientations which match the vast one which is, to my mind, general now in the race. I mean, life has to be sloughed: has to be faced: to be rejected: then accepted on new terms with rapture. And so on, and so on; till you are 40, when the only problem is how to grasp it tighter and tighter to you, so quick it seems to slip, and so infinitely desirable is it [p. 598].

Neugarten (1970), a student of the life cycle and people's ways of coping with the changes time brings about, has pointed out that people who reach middle age tend to restructure their life "in terms of time left to live rather than time since birth. Not only is there a reversal of directionality but also an awareness that time is finite" [p. 78]. As one of Neugarten's subjects said: "All of us figure backward from the end. . . and estimate how much time we can expect" (p. 78).

This change in time perspective is related to the changed perception of death as a possibility that becomes more personal, more probable, more real than it used to be in youth, when death had a rather remote, magical or accidental quality. In aging, one is more concerned about illness, the probability of getting cancer, or just the certainty of getting old—in fact, all hidden versions of the fear of death.

Time and Old Age

After a certain age, the feeling that time goes too fast becomes nearly universal. As William James (1890) put it: "The same space of time seems shorter as we grow older" (p. 284). This phenomenon obtains particularly for longer periods of time—days, seasons, years—and less for short durations. It is indeed striking to recall how time seemed to be infinitely long in childhood, how slowly time passed when waiting for something, even when it was only a matter of waiting for a few minutes.

It has been suggested (Hoagland, 1966) that fever, and in general an increase in the rate of metabolic change in the body and brain, causes subjective time to pass more slowly; and vice-versa, a lower body temperature slows down one's inner sense of time, making time seem to pass faster. Thus changes in the organism's metabolic rate may act like an internal clock and somehow make one perceive the world around as moving faster or slower, faster with age when the metabolic rate slows down, throwing one's internal time out of pace with that of the world. But experimental support of this hypothesis is equivocal.

The way people feel about themselves and the passage of time after a certain age may be compared to the experience of individuals who, voluntarily or not, find themselves under conditions of sensory deprivation. Contrary to the normal impression that "unfilled time" passes very slowly, subjects under conditions that do not allow any external stimulation or change in sensory stimulation (for example, being suspended inside a water tank) report shorter experiential durations than the objective time covering the period of reduced sensory input (Hebb, 1949; Ruff et al., 1961; Zubek, 1969). Similar results were obtained with experiments in which people subjected themselves to isolation in dark subterranean caves for a prolonged period of time (Siffre, 1964, 1975). Deprived of all reliable opportunities to count the passage of time, without any obvious clues to the diurnal cycle other than their own sleep-wakefulness rhythm, people invariably underestimate the passage of time, the duration of their experience.[2]

It is worth noting that, while in isolation, Siffre (1964) experienced other symptoms associated with old age. He reported a progressive impairment of short-term memory—"a fragility of memory," as he put it—and a disturbance in his ability to concentrate on intellectual work by what he described as short periods of absentmindedness. He also suffered from attacks of anxiety, which at times reached the point of despair. He was afraid he would not come out alive and took a pessimistic view of the physical changes he noticed in his face and body, interpreting them as evidence of rapid aging. Finally, he experienced the temptation to justify his suffering not on the basis of the scientific merit of his undertaking, but in terms of some transcendental truth or some superior external power.

In general, as one grows older, there is a strong tendency to underestimate the overall duration of one's experience; at the same time, one feels that the present is slow, monotonous or indifferent, and the future constricted, limited and limiting rather than short or long; and one feels impatient, prone to anxious or depressive concerns about decay and death. As an analyst, I am

[2] A professional geologist in his early twenties Siffre (1964), volunteered to live in a cave beneath a glacier in the Alps for two months. When told that his time was up, he was surprised. He had descended into the cave on July 16, and when taken out on September 14, two days ahead of the planned date, he believed it was only August 20. He had underestimated the length of his experience by 25 days or almost by half its actual length. In a repeat experiment ten years later in Texas, keeping track of his sleep-wake cycles led Siffre (1975) to the erroneous conclusion that he had spent close to 30 days less—out of a total of six months—alone underground. Although his diurnal cycles varied from about 18 to 52 hours, each seemed like a normal day to him. In this and other similar studies, the person in isolation finds himself settling spontaneously onto a 48-hour circadian time consisting of 14 hours sleep and 34 hours wakefulness—twice the normal circadian time.

On the other hand, it has been found that the circadian activity of certain animals becomes progressively shorter with age, suggesting a decay of the biological clock responsible for circadian organization. In reporting such a finding, Pittendrigh and Daan (1974) allude to the common observation that as one grows older time seems to pass faster.

inclined to explain all this as a result of realignment of one's motivational balance and ego structure, to begin with a change in the relative strength of one's instinctual drives and drive derivatives that comes with age.

As one grows older, one's drives tone down, becoming less pressing, less urgent; hence, one becomes less prone to enjoy activities that provide immediate, contact-dependent sensory gratification. As primary wishes become relatively unimportant, it is easier to wait for their fulfillment; and their fulfillment, the experience of pleasure that goes along with their discharge, is in turn not likely to be as novel and exciting an event as in earlier years, when the urgency to repeat such an experience made time appear long and slow-moving. The stronger the wish, the more impatient one gets with the flow of time, which experientially represents the vehicle of getting to the sources of gratification, the means of achieving wish fulfillment. The stronger the wish, the slower the flow of time appears to be; conversely, as the wish loses its urgency—and as long as there is still a wish—time appears to move faster.

The feeling that time goes fast is related to one more factor, something likely to supercede the influence of the id: the internal pressure from one's own superego, the wish to be creative, productive, useful. The older one gets, the more one is likely to depend on activities of an intellectual, professional, or social nature—activities whose motivational origin lies closer to one's superego than id. Affectivity, coming under the dominance of the superego, becomes less intense and pervasive; time becomes less important as an immediate experience and more important as an accomplishment, as a means to an end. Time acquires the value of a product, becomes literally money, productive thinking, opportunity for more planning, more meetings, more sales, more profits, more accomplishments. Time becomes crowded with sublimated content; which, gratifying as it may be morally and intellectually, is only a substitute for the basic, instinctual drives that still seek gratification, yet less directly, with much less urgency than before. As our perspective becomes shorter with the realization that our actuarial future is proportionately smaller, we are concerned about our ability to accomplish enough or, to put it in ego-psychological terms, to satisfy our superego within the remaining lifetime, before it is too late.

Near the end of an eventful and creative life, the poet, novelist, and one-time politician Kazantzakis (1961) confided to his wife, Helen:

> I feel like doing what Bergson says—going to the street corner and holding out my hand to start begging from the passers-by: "Alms, brothers! A quarter of an hour from each of you." Oh, for a little time, just enough to let me finish my work [p. 10].

We grow old but do not feel any older, forgetful at times and breathless when climbing stairs—so long as we do not think of time or death. Weakened as it may become, our id remains timeless, and with it the youthful wish to go on living forever. As Sartre said at seventy:

> I only feel old age through my half-blindness—which is an accident. I could have others—and through the nearness of death, which is absolutely undeniable. Not that I think about it, I never think about it; but I know that it is coming. And then: You know, there was a time when I believed I was immortal, until I was about 30. But now I know that I am very mortal, without ever thinking of death [Sartre and Contat, 1975, p. 14].

As we age, we all tend to become philosophers of time, to comment more and more on life from the perspective of changing times. Time moves and carries us along with it, but not all of ourselves. For the self that we value and remember as our own with joy and pride remains more and more behind; the aspirations, the plans, the illusions become crowded into an ever-shrinking space. As we grow older, time seems to move faster; we wish to catch up with time, to experience what we missed in earlier years, to make up for old mistakes, to complete cherished projects, to renew ourselves into our children's children; until, gradually, we settle on the notion that all has not yet ended, that as long as we keep breathing and our heart endures, there may still be time.

EPILOGUE

Time is man's own sense, the sense of mortals, what distinguishes us from animals and gods. Time and timelessness and all the varieties of temporal experience inbetween have intrigued the human mind ever since we discovered the secret of existence: transience, death.

The idea of transience makes us all feel sad, unless already in pain or driven by an urgent want; then the idea of transience gives us solace. The idea of transience—time passing—makes us sad because, as Freud (1916) saw it, we need to mourn the loss of what we love: beauty, goodness, and the people to whom we become attached before we are able to reinvest our interest and love in something or someone new and as transient. The alternative to sadness and mourning is fear, the fear of death, the passing away of those we love and our own personal existence.

The fear of death and the fear of time are one and the same. Without the fear of death there would be no fear of time. Without the concept of time there would be no concept of death. Time symbolizes death and many other things as well, things essential to our existence and survival, beginning with mother and father, their demands and expectations and control of us through our superego.

To control time is to control oneself and others. The effort to control time by wasting it, by procrastinating, by missing one's analytic appointment, by abruptly terminating or prolonging indefinitely a summer love affair, becomes the effort of controlling one's own potential destructiveness and the fear of rejection, the fear of being left alone, helpless and hopeless. The ultimate defense is to deny time, to live in the present, or to believe in life after death, timeless eternity, a land where there shall be time no longer.

REFERENCES

Abraham, K., (1911), Notes on the psychoanalytic investigation and treatment of manic-depressive insanity and allied conditions. In: *Selected Papers on Psychoanalysis.* New York: Basic Books, 1953, pp. 137–156.

______ (1923), Contributions to the theory of the anal character. In: *Selected Papers of Karl Abraham.* New York: Basic Books, 1953, pp. 370–392.

Alexiou, L. (1956), *Axehastoi Kairoi: Chroniko tis Niotis* [Unforgettable Times: A Chronicle of Youth]. Athens: Bibliop. "Estias."

Alexiou, E. (1966), *Ya Na Ghinei Meghalos* [In Order to Become Great]. Athens: Dorikos.

Allport, G. W. (1955), *Becoming: Basic Considerations for a Psychology of Personality.* New Haven: Yale University Press.

Ames, L. S. (1946), The development of the sense of time in the young child. *J. Genet. Psychol.,* 68:97–125.

Anderson, E. W. (1938), A clinical study of states of "ecstasy" occurring in affective disorders. *J. Neurol. Psychiat.,* 1:1-20.

Anliker, J. (1963), Variations in alpha voltage of electroencephalogram and time perception. *Science,* 140:1307.

Arieti, S. (1947), The processes of expectation and anticipation. Their genetic development, neural basis, and role in psychopathology. *J. Nerv. Ment. Dis.,* 106:471–481.

______ (1955), *Interpretation of Schizophrenia* (2nd ed.). New York: Basic Books, 1974.

Aristotle (384–322 B.C.), *Ethics.* Trans. J. A. K. Thomson. London: Allen & Unwin.

Arlow, J. A. (1966), Depersonalization and derealization. In: R. M. Loewenstein et al., Eds. *Psychoanalysis—A General Psychology.* New York: International Universities Press, pp. 456–478.

Aronson, H. , Silverstein, A. B., & Klee, G. B. (1959), The influence of lysergic acid diethylamide (LSD-25) on subjective time. *Arch. Gen. Psychiat.,* 1:469–472.

Augustinus, A., St. (*ca.* 398), *Confessions and Enchiridion.* Trans. A. C. Outler. Philadelphia: Westminister Press, 1955.

Axelrod, J. (1974), The pineal gland: a neurochemical transducer. *Science,* 184:1341–1348.

Bain, A. (1859), *The Emotions and the Will.* New York: Appleton, 1876.

Balis, G. U. (1974), The use of psychomimetic and related consciousness-altering drugs. In: *American Handbook of Psychiatry* (2nd ed.), Vol. 3, Ed. S. Arieti. New York: Basic Books, pp. 404–446.

Barr, H. L., Langs, R. J., Holt, R. R., Goldberger, L., & Klein, G. S. (1972), *LSD: Personality and Experience.* New York: Wiley-Interscience.

Baudelaire, C. (1860), Les Paradis Artificiels. In: Y. G. Le Dantec, Ed., *Oeuvres Completes de Baudelaire.* Paris: Gallimard, 1961, pp. 345–462.

Bender, L. (1950), Anxiety in disturbed children. In: P. H. Hoch & J. Zubin, Eds., *Anxiety.* New York: Grune & Stratton, pp. 119–139.

Benjamin, J. D. (1961), Some developmental observations relating to the theory of anxiety. *J. Amer. Psychoanal. Assn.,* 9:652–668.

Bergler, E., & Róheim, G. (1946), Psychology of time perception. *Psychoanal. Quart.,* 15:190–206.

Bergson, H. (1889), *Time and Free Will. An Essay on The Immediate Data of Consciousness.* Trans. F. L. Pogson. London: Allen, 1913.

——— (1907), *Creative Evolution.* Trans. A. Mitchell. London: Macmillan, 1911.

——— (1922), *Duration and Simultaneity.* Trans. L. Jacobson. New York: Bobbs-Merrill, 1965.

Bettelheim, B. (1943), Individual and mass behavior in extreme situations. *J. Abnorm. Soc. Psychol.,* 38:417–452.

Bibring, E. (1953), The mechanism of depression. In: P. Greenacre, Ed., *Affective Disorders: Psychoanalytic Contributions to Their Studies.* New York: International Universities Press, pp. 13–48, 137–156.

Binkley, S., Riebman, J. B., & Reilly, K. B. (1977), Timekeeping by the pineal gland. *Science,* 197:1181–1183.

Binswanger, L. (1958), The case of Ellen West; an anthropological-clinical study. In: R. May, E. Angel, and H. F. Ellenberger, Eds., *Existence.* New York: Basic Books.

Bleuler, E. P. (1911), *Dementia Praecox or The Group of Schizophrenias.* Trans. J. Zinkin. New York: International Universities Press, 1950.

——— (1916), *Textbook of Psychiatry.* Trans. A. A. Brill. New York: Macmillan, 1924.

Boisen, A. T. (1936), *The Exploration of the Inner World.* New York: Harper.

Bonaparte, M. (1940), Time and the unconscious. *Internat. J. Psycho-Anal.,* 21:427–468.

Boring, E. G. (1933), *The Physical Dimensions of Consciousness.* New York & London: Century.

Bouchard, R. (1926), *Sur l'Evolution du Temps dans Certains Troubles Mentaux.* Paris: Vigot Frères.

Brenner, C. (1979), Components of psychic conflict and its consequences in mental life. *Psychoanal. Quart.,* 48:547–567.

Breuer, J., & Freud, S. (1893–1895), Studies on hysteria. *Standard Edition,* 2:1–335. London: Hogarth Press, 1955.

Bromberg, W. (1934), Marijuana intoxication. *Amer. J. Psychiat.,* 91:303–330.

Brunswik, E. (1957), Scope and aspects of the cognitive problem. In: *Cognition: The Colorado Symposium.* Cambridge: Harvard University Press.

Bull, N. (1951), *The Attitude Theory of Emotion.* New York: Nervous and Mental Disease Monographs.

Cannon, W. B. (1915), *Bodily Changes in Pain, Hunger, Fear and Rage.* New York: Appleton.

——— (1929), *Bodily Changes in Pain, Hunger, Fear and Rage* (2nd ed.). Boston: Charles T. Branford, 1953.

Cantril, H., & Hunt, W. A. (1932), Emotional effects produced by injection of adrenalin. *Amer. J. Psychol.,* 44:300–307.

Ciompi, L. (1961), Abnormal time sense in a schizophrenic female. *Psychiat. & Neurol.,* 142:100–121.

Clark, L. D., Hughes, R., & Nakashima, E. N. (1970), Behavioral effects of marijuana. *Arch. Gen. Psychiat.,* 23:193–198.

Clay, E. R. (1882), *The Alternative. A Study in Psychology.* London.

Colarusso, C. (1979), The development of time sense—from birth to object constancy. *Internat. J. Psycho-Anal.*, 60:243–251.
Cottle, T. J., & Klineberg, S. L. (1974), *The Present of Things Future.* New York: Free Press.
Darwin, C. (1872), *The Expression of the Emotions in Man and Animals.* Chicago: University of Chicago Press, 1965.
Décarie, T. G. (1976), Affect development and cognition in a Piagetian context. In: M. Lewis & L. A. Rosenblum, Eds., *The Development of Affect.* New York: Plenum, pp. 183–204.
Deikman, A. J. (1963), Experimental meditation. *J. Nerv. Ment. Dis.*, 136:329–373.
______ (1966), De-automatization and the mystical experience. *Psychiat.*, 29:324–338.
DeQuincey, T. (1822), *The Confessions of An English Opium-Eater.* New York: Dutton, 1907.
de Romilly, J. (1968), *Time in Greek Tragedy.* Ithaca, N.Y.: Cornell University Press.
Dewey, J. (1886), *Psychology.* New York: Harper, 1898.
Dickinson, E. (1890), *The Complete Poems.* Ed. T. H. Johnson. Boston & Toronto: Little, Brown, 1960.
Dilling, L. A., & Rabin, A. I. (1967), Temporal experience in depressive states and schizophrenia. *J. Consult. Psychol.*, 31:604–608.
Dimond, S. J. (1980), *Neuropsychology.* London: Butterworths.
Dodds, E. R. (1951), *The Greeks and the Irrational.* Berkeley & Los Angeles: University of California Press.
Doob, L. W. (1971), Patterning of Time. New Haven: Yale University Press.
Dooley, L. (1941), The concept of time in defense of ego integrity. *Psychiat.*, 4:13–23.
Dostoevski, F. M. (1868), *The Idiot.* Trans. C. Garnett. New York: Modern Library, 1935.
Dubois, F. S. (1955), The sense of time and its relation to psychiatric illness. *Amer. J. Psychiat.*, 111:46–51.
Ebin, D., Ed. (1961), *The Drug Experience.* New York: Orion Press.
Eccles, J. C., Ed. (1966), *Brain and Conscious Experience.* New York: Springer.
Efron, R. (1963), Temporal perception, aphasia, and *déjà vu. Brain,* 86:403–424.
Ehrensing, R. H., Stokes, P. E., Pick, G. R., Goldstone, S. & Lhamon, W. T. (1970), Effect of alcohol on auditory and visual time perception. *Quart. J. Stud. Alcohol,* 31:851–860.
Ehrentheil, O. F., & Jenney, P. B. (1960), Does time stand still for some psychotics? *Arch. Gen. Psychiat.*, 3:1–3.
Eisendrath, C. R. (1971), *The Unifying Moment: The Psychological Philosophy of William James and Alfred North Whitehead.* Cambridge, Mass.: Harvard University Press.
Eissler, K. R. (1952), Time experience and the mechanisms of isolation. *Psychoanal. Rev.*, 39:1–21.
______ (1955), *The Psychiatrist and the Dying Patient.* New York: International Universities Press.
Epstein, S. (1967), Toward a unified theory of anxiety. In: B. A. Mahel, Ed., *Progress in Experimental Personality Research,* Vol. 4. New York: Academic Press, pp. 1–89.
Erikson, E. H. (1950), *Childhood and Society* (2nd ed.). New York: Norton, 1963.
______ (1954), The dream specimen of psychoanalysis. *J. Amer. Psychoanal. Assn.*, 2:5–56.
______ (1956), The problem of ego identity. *J. Amer. Psychoanal. Assn.*, 4:56–121.
______ (1959), *Identity and the Life Style* [Psychological Issues, Monogr. 1]. New York: International Universities Press.

Ey, H. (1969), Disorders of consciousness in psychiatry. In: P. J. Vinken & G. W. Bruyn, Eds., *Handbook of Clinical Neurology—Disorders of Higher Nervous Activity.* Amsterdam: North-Holland Publishing Co., Vol. 3, pp. 112-136.

Fairbairn, W. R. D. (1941), A revised psychopathology of the psychoses and psychoneuroses. *Internat. J. Psycho-Anal.,* 22:250-279.

——— (1952), *An Object-Relations Theory of Personality.* New York: Basic Books.

Fanourakis, G. (1958), Ermineutikes paratiriseis ston kapetan Mihale [Interpretive observations on Capetan Michalis]. *Kainouria Epoche,* 3:184-199.

Fenichel, O. (1945), *The Psychoanalytic Theory of Neurosis.* New York: Norton.

——— (1953), On the psychology of boredom. In: *The Collected Papers of Otto Fenichel.* New York: Norton, pp. 292-302.

Fingarette, H. (1958), The ego and mystic selflessness. *Psychoanal. Rev.,* 45:5-40.

Fischer, R. (1971), A cartography of the ecstatic and meditative states. *Science,* 174:897-904.

Flugel, J. D. (1955), *Studies in Feeling and Desire.* London: Duckworth.

Fraiberg, S. (1969), Libidinal object constancy and mental representation. *Psychoanal. Study Child.,* 24:9-47.

Fraisse, P. (1963), *The Psychology of Time.* New York: Harper & Row.

Fraser, J. T. (1972), The concept of time in Western thought. *Main Currents in Modern Thought,* 28:115-124.

——— Ed. (1966), *The Voices of Time.* New York: Braziller.

——— (1975), *Of Time, Passion, and Knowledge.* New York: Braziller.

Freedman, D. X. (1968), On the use and abuse of LSD. *Arch. Gen. Psychiat.,* 18: 330-347.

Freud, A. (1936), *The Ego and the Mechanisms of Defense.* New York: International Universities Press, 1966.

——— (1965), *Normality and Pathology in Childhood.* New York: International Universities Press.

Freud, S. (1895), Project for a scientific psychology. *Standard Edition,* 1:295-387. London: Hogarth Press, 1966.

——— (1900), The interpretation of dreams. *Standard Edition,* 5:1-621. London: Hogarth Press, 1962.

——— (1901), The psychopathology of everyday life. *Standard Edition,* 6:1-310. London: Hogarth Press, 1960.

——— (1909), Notes upon a case of obsessional neurosis. *Standard Edition,* 10: 151-249. London: Hogarth Press, 1962.

——— (1913), Observations and examples from analytic practice. *Standard Edition,* 13:193-198. London: Hogarth Press, 1962.

——— (1914), On narcissism: an introduction. *Standard Edition,* 14:67-102. London: Hogarth Press, 1957.

——— (1915), The unconscious. *Standard Edition,* 14:159-215. London: Hogarth Press, 1957.

——— (1916), On transcience. *Standard Edition,* 14:305-307. London: Hogarth Press, 1957.

——— (1917), Mourning and melancholia. *Standard Edition,* 14:237-260. London: Hogarth Press, 1957.

——— (1920), Beyond the pleasure principle. *Standard Edition,* 18:3-64. London: Hogarth Press, 1955.

——— (1923), The ego and the id. *Standard Edition,* 19:1-59. London: Hogarth Press, 1961.

——— (1925a), Negation. *Standard Edition,* 19:235–239. London: Hogarth Press, 1962.

——— (1925b), A note upon the "Mystic Writing-Pad." *Standard Edition,* 19:227–232. London: Hogarth Press, 1962.

——— (1926), Inhibitions, symptoms and anxiety. *Standard Edition,* 20:77–172. London: Hogarth Press, 1959.

——— (1928), A religious experience. *Standard Edition,* 21:167–172. London: Hogarth Press, 1961.

——— (1930), Civilization and its discontents. *Standard Edition,* 21:117–139. London: Hogarth Press, 1961.

——— (1933), New introductory lectures on psycho-analysis. *Standard Edition,* 22:5–182. London: Hogarth Press, 1964.

——— (1936), A disturbance of memory on the Acropolis. *Standard Edition,* 22: 236–248. London: Hogarth Press, 1964.

——— (1938), Findings, ideas, problems. *Standard Edition,* 23:299–300. London: Hogarth Press, 1964.

Friar, K. (1960), I pneumatiki askisi tou Nikou Kazantzaki [The spiritual exercise of Nikos Kazantzakis]. *New Estia,* 33(779):69–85.

Gifford, S. (1960), Sleep, time and the early ego. Comments on the development of the 24-hour sleep-wakefulness pattern as a precursor of ego functioning. *J. Amer. Psychoanal. Assn.,* 8:5–42.

Gill, M., & Brenman, M. (1959), *Hypnosis and Related States: Psychoanalytic Studies in Regression.* New York: International Universities Press.

Goldstein, K. (1939), *The Organism.* New York: American Book Co.

Goldstone, S., Boardman, W. I., & Lhamon, W. T. (1958), Effects of quinalbarbitone, dextroamphetamine and placebo on apparent time. *Brit. J. Psychol.,* 49:324–328.

——— & Kirkham, J. (1968), The effects of secobarbital and dextroamphetamine upon time judgment; intersensory factors. *Psychopharmacologia,* 13:65–73.

Gooddy, W. (1969), Disorders of the time sense. In: P. J. Vinken & G. W. Bruyn, Eds., *Handbook of Clinical Neurology—Disorders of Higher Nervous Activity.* Amsterdam: North-Holland Publishing Co., Vol. 13, pp. 229–250.

Gorman, B. S., & Wessman, A. E., Eds. (1977), *The Personal Experience of Time.* New York: Plenum.

Greenson, R. R. (1953), On boredom. *J. Amer. Psychoanal. Assn.,* 1:7–21.

——— (1971), A dream while drowning. In: J. B. McCevitt & C. F. Settlage, Eds., *Separation-Individuation. Essays in Honor of Margaret S. Mahler.* New York: International Universities Press, pp. 377–384.

Greyson, B., & Stevenson, I. (1980), The phenomenology of near-death experiences. *Amer. J. Psychiat.,* 137:1193–1196.

Grinberg, L. (1964), Two kinds of guilt—their relations with normal and pathological aspects of mourning. *Internat. J. Psycho-Anal.,* 45:336–371.

——— (1977), An approach to the understanding of borderline disorders. In: P. Hartocollis, Ed., *Borderline Personality Disorders: The Concept, the Syndrome, the Patient.* New York: International Universities Press, 1977, pp. 123–141.

Grinker, R., Werble, B., & Drye, R. (1968), *The Borderline Syndrome.* New York: Basic Books.

Gross, A. (1949), Sense of time in dreams. *Psychoanal. Quart.,* 18:446–470.

Gruber, H. D., & Voneche, J. D., Eds. (1977), *The Essential Piaget: An Interpretive Reference and Guide.* New York: Basic Books.

Guttman, E., & MacClay, W. C. (1936), Mescalin and depersonalization. *J. Neurol. Psychopathol.*, 16:193–212.

Hartmann, H. (1950), Comments on the psychoanalytic theory of the ego. *Psychoanal. Study Child.*, 5:74–96.

______ (1964), *Essays on Ego Psychology. Selected Problems in Psychoanalytic Theory.* New York: International Universities Press.

______ Kris, E., & Loewenstein, R. M. (1946), Comments on the formation of psychic structure. *Psychoanal. Study Child,* 2:11–38.

Hartocollis, P. (1962), Drunkenness and suggestion: an experiment with intravenous alcohol. *Quart. J. Stud. Alcohol,* 23:376–389.

Hebb, D. O. (1949), *Organization of Behavior.* New York: Wiley.

Heidegger, M. (1927), *Being and Time.* Trans. J. Macquarrie & E. Robison. New York: Harper & Row, 1962.

Hoagland, H. (1933), The physiological control of judgments of duration: evidence for a chemical clock. *J. Gen. Psychol.*, 9:267–287.

______ (1966), Some biochemical considerations of time. In: J. T. Fraser, Ed., *The Voices of Time.* New York: Braziller, pp. 312–329.

Hoch, P. H., Cattell, J. P., & Pennes, H. H. (1952), Effects of mescaline and lysergic acid (d-LSD-25). *Amer. J. Psychiat.*, 108:579–584.

Hoffmann, A. (1959), Psychotomimetic drugs. Chemical and pharmacological aspects. *Acta Physiol. Pharmacol. Neerlandica.*, 8:240–258.

Holubar, J., & Machek, J. (1962), Time sense and epileptic EEG activity. *Epilepsia* (Amst.) 3(3–4):323.

Huxley, A. (1954), *The Doors of Perception.* New York: Harper & Row.

Inhelder, B., & Piaget, J. (1955), *De la Logique de l'Enfant à la Logique de l'Adolescent.* Paris: Press Universitaires de France.

Isaacs, S. (1948), The nature and function of phantasy. *Internat. J. Psycho-Anal.*, 29:73–97.

Israeli, N. (1936), *Abnormal Personality and Time.* New York: Science Press.

Jackson, J. H. (1874-1876), On the scientific and empirical investigation of epilepsies. In: J. Taylor, Ed., *Selected Writings of John Hughlings Jackson,* Vol. 1. New York: Basic Books, 1958, pp. 162–273.

______ (1880), On right- or left-sided spasm at the outset of epileptic paroxysms, and on crude sensation warnings, and elaborate mental states. In: J. Taylor, Ed., *Selected Writings of John Hughlings Jackson,* Vol. 1. New York: Basic Books, 1958, pp. 308-317.

______ (1888), On a particular variety of epilepsy ("intellectual aura"), one case with symptoms of organic disease. In: J. Taylor, Ed., *Selected Writings of John Hughlings Jackson,* Vol. 1. New York: Basic Books, 1958, pp. 385–405.

Jacobson, E. (1953), The affects and their pleasure-unpleasure qualities in relation to the psychic discharge processes. In: R. M. Loewenstein, Ed., *Drives, Affects, Behavior,* Vol. 1. New York: International Universities Press, pp. 38–66.

______ (1957), Normal and pathological moods: their nature and functions. *Psychoanal. Study Child.*, 12:73–113.

______ (1964), *The Self and the Object World.* New York: International Universities Press.

James, W. (1884), What is an emotion? *Mind,* 9:188–205.

______ (1890), *The Principles of Psychology,* Vol. 1. New York: Holt.

______ (1902), *The Varieties of Religious Experience.* New York: Random House, 1929.

Janet, P. (1928), *L'Evolution de la Mémoire et de la Notion du Temps.* Paris: Chahine.

Jaspers, K. (1959), *General Psychopathology.* Trans. J. Hoenic & M. W. Hamilton. Chicago: The University of Chicago Press, 1968.

Joffe, W. G., & Sandler, J. (1965), Notes on pain, depression, and individuation. *Psychoanal. Study Child.,* 20:394–424.

Jones, E. (1918), Anal-erotic character traits. In: *Papers on Psycho-Analysis.* London: Baillière, Tindal & Cox. (5th ed.), 1950, pp. 413–437.

——— (1953-1957), *The Life and Work of Sigmund Freud,* 3 Vols. New York: Basic Books.

Jouvenel, R. de (1959), Eis mnimin Kazantzaki (In memory of Kazantzakis]. *Nea Estia,* 33(779):228–243.

Kagan, J. (1970), Attention and psychological change in the young child. *Science,* 170:826–832.

Katz, M. M., Waskow, I. E., & Olsson, J. (1968), Characterizing the psychological state produced by LSD. *J. Abnorm. Psychol.,* 73(1):1–14.

Kazantzakis, G. (1957), *Anthropoi kai Yperanthropoi [Men and Supermen].* Athens: Pyxida.

Kazantzakis, H. (1968), *Nikos Kazantzakis: A Biography.* New York: Simon & Schuster.

Kazantzakis, N. (1927), *The Saviours of God—Spiritual Exercises.* Trans. K. Friar. New York: Simon & Schuster, 1960.

——— (1938), *The Odyssey: A Modern Sequel.* Trans. K. Friar. New York: Simon & Schuster, 1958.

——— (1946), *Zorba the Greek.* Trans. C. Wildman. New York: Simon & Schuster, 1952.

——— (1948), *The Greek Passion.* Trans. J. Griffin. New York: Simon & Schuster, 1954.

——— (1950), *Freedom or Death.* Trans. J. Griffin. New York: Simon & Schuster, 1956.

——— (1955), *The Last Temptation of Christ.* Trans. P. A. Bien. New York: Simon & Schuster, 1960.

——— (1958), *Epistoles pros ti Galatia* [Letters to Galatea]. Athens: Diphros.

——— (1961), *Report to Greco.* Trans. P. A. Bien. New York: Simon & Schuster, 1965.

——— (1971), *Symposium.* New York: Minerva Press, 1974. Trans. Th. & Th. Vasils.

Kenna, J. D., & Sedman, G. (1964), The subjective experience of time during lysergic acid diethylamide (LSD-25) intoxication. *Psychopharmacologia,* 5:280–288.

Kernberg, O. (1975), *Borderline Conditions and Pathological Narcissism.* New York: Aronson.

——— (1976), *Object Relations Theory and Clinical Psychoanalysis.* New York: Aronson.

Kierkegaard, S. (1844), *The Concept of Dread.* Princeton: Princeton University Press, 1944.

Klee, G. D. (1963), Lysergic acid diethylamide (LSD-25) and ego functions. *Arch. Gen. Psychiat.,* 8:461–474.

Klein, M. (1932), *The Psycho-Analysis of Children.* New York: Norton.

——— (1935), A contribution to the psychogenesis of manic-depressive states. *Internat. J. Psycho-Anal.,* 16:145–174.

——— (1946), Notes on some schizoid mechanisms. *Internat. J. Psycho-Anal.,* 27:99–110.

Klüver, H. (1928), *Mescaline [Mescal] The "Divine" Plant and Its Psychological Effects.* London: Kegan Paul, Trench, Trubner.

Kohut, H. (1971), *The Analysis of the Self.* New York: International Universities Press, 1971.

Kraines, S. H. (1957), *Mental Depressions and Their Treatment.* New York: Macmillan.

Krus, D. M., Wapner, S., Freeman, H., & Casey, T. M. (1963), Differential behavioral responsivity to LSD-25: study in normal and schizophrenic adults. *Arch. Gen. Psychiat.,* 8:557-563.

Kubie, L. S. (1941), A physiological approach to the concept of anxiety. *Psychosom. Med.,* 3:263-276.

______ (1954), Psychiatric and psychoanalytic considerations of the problem of consciousness. In: J. F. Delafresnaye, Ed., *Brain Mechanisms and Consciousness.* Oxford: Blackwell Scientific Publications, pp. 444-469.

______ (1963), The central affective potential and its trigger mechanisms. In: H. S. Gaskill, Ed., *Counterpoint: Libidinal Object and Subject.* New York: International Universities Press, pp. 106-120.

Leeds, M. (1944), One form of paramnesia: the illustration of *déjà vu. J. Amer. Soc. Psychic Res.,* 38:24-42.

Leeper, R. W. (1970), The motivational and perceptual properties of emotions as indicating their fundamental character and role. In: M. B. Arnold, Ed., *Feelings and Emotions.* New York: Academic Press, pp. 151-168.

Lehmann, H. E. (1967), Time and psychopathology. *Annals Acad. Sci.,* 138:798-821.

Levin, D. C. (1969), The self: a contribution to its place in theory and technique. *Internat. J. Psycho-Anal.,* 50:41-51.

Lewin, B. D. (1950), *The Psychoanalysis of Elation.* New York: Norton.

______ (1952), Phobic symptoms and dream interpretation. *Psychoanal. Quart.,* 21:295-322.

Loewald, H. W. (1962), Superego and time. *Internat. J. Psycho-Anal.,* 43:264-268.

______ (1972), The experience of time. *Psychoanal. Study Child.,* 27:401-410.

______ (1978), *Psychoanalysis and the History of the Individual.* New Haven & London: Yale University Press.

Lubin, A. J. (1977), From Augustine to Einstein: thoughts on mysticism, creativity, and identity. *Dialogue: A Journal of Psychoanalytic Perspectives,* 1:31-44.

Mahler, M. D. (1966), Notes on the development of basic moods—the depressive affect. In: R. M. Loewenstein et al., Eds., *Psychoanalysis—A General Psychology.* New York: International Universities Press, pp. 152-168.

______ Pine, F., & Bergman, A. (1975), *The Psychological Birth of the Human Infant—Symbiosis and Individuation.* New York: Basic Books.

Malraux, A. (1974), *Lazarus.* Trans. T. Kilmartin. New York: Holt, Rinehart & Winston, 1977.

Mann, T. (1924), *The Magic Mountain.* Trans. H. T. Lowe-Porter. New York: Knopf, 1947.

Maranon, G. (1924), Contribution a l'étude de l'action émotive de l'adrenaline. *Rev. Franç. Endocrinol.,* 2:301-325.

Masler, E. G. (1973), The subjective perception of two aspects of time: duration and timelessness. *Internat. J. Psycho-Anal.,* 54:425-429.

McDougall, W. (1923), *Outline of Psychology.* New York: Scribner.

Meerloo, A. M. (1948), Father time. An analysis of subjective conceptions of time. *Psychiat. Quart.,* 22:587-608.

______ (1954), *The Two Faces of Man. Two Studies on the Sense of Time and on Ambivalence.* New York: International Universities Press.

Melges, F. T., & Fongerousse, C. E. (1966), Time sense, emotions, and acute mental illness. *J. Psychiat. Res.,* 4:127-140.

______ & Freeman, A. M. (1977), Temporal disorganization and inner-outer confusion in acute mental illness. *Amer. J. Psychiat.,* 132:874-877.

______ Tinklenberg, J. R., Hollister, L. E. & Gillespie, H. K. (1971), Marijuana and the temporal span of awareness. *Arch. Gen. Psychiat.,* 24:564–567.

Meyer-Gross, W. (1935), On depersonalization. *Brit. J. Med. Psychol.,* 15:103-126.

Mezey, A., & Cohen, S. (1961), The effect of depressive illness on time judgment and time experience. *J. Neurol. Neurosurg. Psychiat.,* 24:269–270.

Minkowski, E. (1933), *Lived Time: Phenomenological and Psychopathological Studies.* Trans. N. Metzel. Evanston: Northwestern University Press, 1970.

Money-Kyrle, R. E. (1955), Psychoanalysis and ethics. In: M. Klein et al., Eds., *New Directions in Psychoanalysis.* London: Tavistock, pp. 421–439.

Morris, J. L. (in press), Time experience and transference. *J. Amer. Psychoanal. Assn.*

Nacht, S., & Racamier, P. C. (1960), Symposium on "depressive illness." II, depressive states. *Internat. J. Psycho-Anal.,* 41:481–496.

Neugarten, B. L. (1970), Adaptation and the life cycle. *J. Geriat. Psychiat.,* 1:21–87.

New York Times (Jan. 30, 1979), p. 10a.

Noyes, R., & Kletti, R. (1976), Depersonalization in the face of life-threatening danger: a description. *Psychiat.,* 39:19–27.

Nunberg, H., & Federn, E., Eds. (1974), *Minutes of the Vienna Psychoanalytic Society,* Vol. 3, 1910–1911. New York: International Universities Press.

Oberndorf, C. P. (1941), Time—its relation to reality and purpose. *Psychoanal. Rev.,* 28:139–155.

O'Brien, E. (1964), *Varieties of Mystic Experience.* New York: Holt, Rinehart & Winston.

Orgel, S. (1965), On time and timelessness. *J. Amer. Psychoanal. Assn.,* 13:102–121.

Ornstein, R. E. (1969), *On the Experience of Time.* Harmondsworth, Middlesex, England: Penguin Books.

Penfield, W. (1959), The interpretive cortex. *Science,* 129:1719–1725.

______ (1975), *The Mystery of the Mind.* Princeton, N.J.: Princeton University Press.

Panel (1972), The experience of time. *J. Amer. Psychoanal. Assn.,* 20:650–667.

Pfaff, D. (1968), Effects of temperature and time of day on time judgments. *J. Exper. Psychol.,* 76:419–422.

Piaget, J. (1936), *The Origins of Intelligence in Children.* Trans. M. Cook. New York: International Universities Press, 1952.

______ (1937), *The Construction of Reality in the Child.* Trans. M. Cook. New York: Basic Books, 1954.

______ (1945), *Play, Dreams and Imitation in Childhood.* Trans. C. Gattegno & F. M. Hodgson. New York: Norton, 1962.

______ (1946), *The Child's Conception of Time.* Trans. A. J. Pomerans. New York: Basic Books, 1969.

______ (1962), The relation of affectivity to intelligence in the mental development of the child. *Bull. Menninger Clin.,* 26:129–137.

______ (1966), Time perception in children. In: Fraser, Ed. (1966), pp. 202–216.

______ (1970), *Structuralism.* New York: Basic Books.

Pittendrigh, C. S., & Daan, S. (1974), Circadian oscillations in rodents: a systematic increase of their frequency with age. *Science,* 186:548–550.

Pollock, G. H. (1970), Anniversary reactions, trauma and mourning. *Psychoanal. Quart.,* 39:347–371.

______ (1971a), On time, death, and immortality. *Psychoanal. Quart.,* 40:435–446.

______ (1971b), On time and anniversaries. In: M. Kanzer, Ed., *The Unconscious Today.* New York: International Universities Press, pp. 233–257.

Popper, K. R., & Eccles, J. C. (1977), *The Self and Its Brain.* New York: Springer.

Prevelakis, P. (1965), *Tetracosia Grammata tou Kazantzaki ston Prevelaki* [Four Hundred

Letters of Kazantzakis to Prevelakis]. Athens: Ekdoseis Elenis N. Kazantzakis.

Pribram, K. (1967), Emotion: steps toward a neuropsychological theory. In: D. C. Glass, Ed., *Neurophysiology and Emotion.* New York: Rockefeller University Press & Russell Sage Foundation, pp. 3-40.

______ (1970), Feelings as monitors. In: M. B. Arnold, Ed., *Feelings and Emotions.* New York: Academic Press, pp. 41-53.

______ (1971), *Languages of the Brain.* Englewood Cliffs, N.J.: Prentice-Hall.

______ & Melges, F. T. (1969), Psychophysiological basis of emotion. In: P. J. Vinken & G. W. Bruyn, Eds., *Handbook of Clinical Neurology—Disorders of Higher Nervous Activity,* Vol. 3. Amsterdam: North-Holland Publishing Co., pp. 316-340.

Pulver, S. E. (1971), Can affects be unconscious? *Internat. J. Psycho-Anal.,* 52:547-554.

Rabin, A. I. (1957), Time estimation of schizophrenics and nonpsychotics. *J. Clin. Psychol.,* 13:88-90.

Rado, S. (1933), The psychoanalysis of pharmacothymia. *Psychoanal. Quart.,* 2:1-23.

Ramzy, I., & Wallerstein, R. S. (1958), Pain, fear and anxiety, a study in their interrelationships. *Psychoanal. Study Child,* 13:147-189.

Rank, O. (1914), *The Double: A Psychoanalytic Study.* Chapel Hill: University of North Carolina Press.

Rapaport, D. (1945), *Diagnostic Psychological Testing,* Vol. I. Chicago: Year Book Publishers.

______ (1953), On the psycho-analytic theory of affects. *Internat. J. Psycho-Anal.,* 34: 177-198.

Reichenbach, H. (1951), *The Rise of Scientific Philosophy.* Berkeley & Los Angeles: University of California.

Richardson, T. F., & Winocur, G. (1967), *Déjà vu* in psychiatric and neurosurgery patients. *Arch. Gen. Psychiat.,* 17:622-625.

Ricoeur, P. (1970), *Freud and Philosophy: An Essay on Interpretation.* Trans. D. Savage. New Haven & London: Yale University Press.

Rosen, D. H., & Hoffman, A. M. (1962), Focal suicide: self-enucleation by two young psychotic individuals. *Amer. J. Psychiat.,* 128:1009-1012.

Ross, N. (1975), Affect as cognition: with observations on the meanings of mystical states. *Internat. Rev. Psycho-Anal.,* 2:79-93.

Ruff, G. E., Levy, E. Z., & Thaler, V. H. (1961), Factors influencing reactions to reduced sensory input. In: P. Solomon et al., Eds., *Sensory Deprivation.* Cambridge, Mass.: Harvard University Press, pp. 72-90.

Russell, B. (1927), *Philosophy.* New York: Norton.

______ (1929), *Mysticism and Logic.* London: Allen & Unwin.

______ (1945), *A History of Western Philosophy.* New York: Simon & Schuster.

______ (1948), *Human Knowledge. Its Scope and Limits.* New York: Simon & Schuster.

______ (1967), *The Autobiography of Bertrand Russell,* Vol. I, 1871-1914. Boston: Atlantic, Little, Brown.

Saint Symeon the New Theologian (1792), Peri ton trion tropon tis proseuchis (about the manners of the prayer). In: *Philocalia,* Vol. 5. Athens: Astir, 1963, pp. 81-89.

Salzman, L. (1953), The psychology of religious and ideological conversion. *Psychiat.,* 16:177-187.

Sandler, J., & Joffe, W. G. (1969), Towards a basic psychoanalytic model. *Internat. J. Psycho-Anal.,* 50:79-90.

Sartre, J. P., & Contat, M. (1975), Sartre at seventy: an interview. *The New York Review of Books,* August 7, 1975, pp. 10-17.

Savage, C. (1955), Variations in ego feeling induced by d-lysergic acid diethylamide (LSD-25). *Psychoanal. Rev.*, 42:1-16.

______ Terrill, J., & Jackson, D. D. (1962), LSD, transcendence, and new beginning. *J. Nerv. Ment. Dis.*, 135:425-439.

Schachtel, E. G. (1959), *Metamorphosis.* New York: Basic Books.

Schachter, S. (1970), The assumption of identity and peripheralist-centralist controversies in motivation and emotion. In: M. B. Arnold, Ed., *Feelings and Emotions.* New York: Academic Press, pp. 111-121.

______ & Singer, J. E. (1962), Cognitive, social and psychological determinants of emotional state. *Psychol. Rev.*, 69:379-399.

Schafer, R. (1964), The clinical analysis of affects. *J. Amer. Psychoanal. Assn.*, 12:275-299.

______ (1976), *A New Language for Psychoanalysis.* New Haven & London: Yale University Press.

Schiffer, I. (1978), *The Trauma of Time.* New York: International Universities Press.

Schilder, P. (1935), *The Image and Appearance of the Human Body.* New York: International Universities Press, 1950.

______ (1936), Psychopathology of time. *J. Nerv. Ment. Dis.*, 83:530-546.

______ (1942), *Mind, Perception and Thought in Their Constructive Aspects.* New York: Columbia University Press.

Schmale, A. H. (1958), Relationship of separation and depression to disease. *Psychosom. Med.*, 20:259-277.

______ (1964), A genetic view of affects: with special reference to the genesis of helplessness and hopelessness. *Psychoanal. Study Child.*, 19:287-310.

Schreber, D. P. (1903), *Memoirs of My Nervous Illness.* Ed. I. Macalpine & R. S. Hunter. London: Dawson, 1955.

Schur, M. (1953), The ego in anxiety. In: R. M. Loewenstein, Ed., *Drives, Affects, Behavior,* Vol. 1. New York: International Universities Press, pp. 67-103.

______ (1969), Affects and cognition. *Internat. J. Psycho-Anal.*, 50:647-653.

______ (1972), *Freud: Living and Dying.* New York: International Universities Press.

Scott, W. C. M. (1948), Some psycho-dynamic aspects of disturbed perception of time. *Brit. J. Med. Psychol.*, 31:111-120.

Searles, H (1965), *Collected Papers on Schizophrenia and Related Subjects.* New York: International Universities Press.

Sedman, G. (1970), Theories of depersonalization: a re-appraisal. *Brit. J. Psychiat.*, 117.1-11.

Siffre, M. (1964), *Beyond Time.* Trans. & ed. H. Briffault. New York: McGraw Hill.

______ (1975), Six months alone in a cave. *Natl. Geog.*, 147:426-435.

Simon, B., & Simon, N. (1972), The pacifist turn: an episode of mystic illumination in the autobiography of Bertrand Russell. *J. Amer. Psychoanal. Assn.*, 20:109-121.

Singer, M. (1977), The experience of emptiness in narcissistic and borderline states: II. The struggle for a sense of self and the potential for suicide. *Internat. Rev. Psycho-Anal.*, 4(4):471-479.

Smith, S. (1977), The golden fantasy: a regressive reaction to separation anxiety. *Internat. J. Psycho-Anal.*, 58:311-324.

Sakolov, E. H. (1960), Neuronal models and the orienting reflex. In M. A. B. Brazier, Ed., *The Central Nervous System and Behavior.* New York: Josiah Macy Jr. Foundation, pp. 187-276.

Spiegel, E., Wycis, H., Orchinik, C., & Freed, H. (1956), Thalamic Chronotaraxis. *Amer. J. Psychiat.*, 113:97.

Spiegel, L. A. (1966), Superego and the function of anticipation with comments on "anticipatory anxiety." In: R. M. Loewenstein et al., Eds., *Psychoanalysis—A General Psychology. Essays in Honor of Heinz Hartmann.* New York: International Universities Press, pp. 315-337.

Spitz, R. A. (1957), *No and Yes. On the Genesis of Human Communication.* New York: International Universities Press.

______ (1972), Bridges. On anticipation, duration, and meaning. *J. Amer. Psychoanal. Assn.,* 20:721-735.

Stace, W. T. (1967), *Against Darkness and Other Essays.* Pittsburgh: University of Pittsburgh Press.

Stekel, W. (1912), Time in its relationship to the neurotic. *Amer. J. Urol. Sexol.,* 14:436-447, 1918.

______ (1927), *Compulsion and Doubt.* Trans. E. Gutheil. London & New York: Peter Nevill, 1950.

______ (1950), The Autobiography of Wilhelm Stekel. New York: Liverright.

Steinberg, H. (1955), Changes in time perception induced by an anesthetic drug. *Brit. J. Psychol.,* 46:273-279.

Stendhal, M. H. B. (1831), *Le Rouge et le Noir.* Paris: Classiques Garnier, 1960.

Sterba, R. (1968), Remarks on mystic states. *Amer. Imago,* 25:77-85.

Stern, W. (1924), *Psychology of Early Childhood Up to the Sixth Year of Age.* Trans. A. Barwell. London: Allen & Unwin.

Straus, E. W. (1947), *Phenomenological Psychology.* Trans. E. Eng. New York: Basic Books, 1966.

Sullivan, H. S. (1954), *The Psychiatric Interview.* New York: Norton.

Tucker, G. J., Harrow, M. & Quinlan, D. (1973), Depersonalization, dysphoria and thought disturbance. *Amer. J. Psychiat.,* 130:702-706.

Underhill, E. (1926), *The Mystics of the Church.* New York: Schoken Books, 1964.

van der Waals, H. G. (1948), The genesis of guilt. In: J. C. Flugel, Ed., *International Congress on Mental Health, Vol. 3: Proceedings of the International Conference on Medical Psychotherapy.* London: Lewis, pp. 4-10.

van Ree, F. (1960), Prediction of syndromes in LSD-25. *Adv. Psychosom. Med.,* 1:209-212.

Waelder, R. (1967), Inhibitions, symptoms, and anxiety forty years later. *Psychoanal. Quart.,* 36:1-36.

Weil, A. T. (1969), In: "Study finds marijuana effects mild." *Psychiat. News,* Feb., p. 25.

______ Zinberg, N. E., & Nelson, J. M. (1968), Clinical and psychological effects of marijuana in man. *Science,* 162:1234-1242.

Werboff, J. (1962), Time judgment as a function of electroencephalographic activity. *Exper. Neurol.,* 6:152.

Werner, H. (1957), *Comparative Psychology of Mental Development,* (rev. ed.) New York: International Universities Press.

______ & Kaplan, B. (1963), *Symbol Formation. An Organismic-Developmental Approach to Language and the Expression of Thought.* New York: Wiley.

Whitehead, A. N. (1927), *Symbolism: Its Meaning and Effect.* New York: MacMillan.

______ (1929), *Process and Reality.* New York: Macmillan.

Whitrow, G. J. (1961), *The Natural Philosophy of Time.* London: Thomas Nelson & Sons.

Williams, D. (1966). Temporal lobe epilepsy. *Brit. Med. J.,* 5501:1439-1442.

Winnicott, D. W. (1958a), *Collected Papers.* New York: Basic Books.

______ (1958b), Psycho-analysis and the sense of guilt. In: J. D. Sutherland, Ed., *Psycho-Analysis and Contemporary Thought.* London: Hogarth Press, pp. 15-32.

______ (1965), *The Maturational Processes and the Facilitating Environment: Studies in the*

Theory of Emotional Development. New York: International Universities Press.

Wittgenstein, L. (1921), *Tractatus Logico-Philosophicus.* Trans. D. F. Pears & B. F. McGuinness. London: Routledge & Kegan Paul, 1974.

——— (1966), *Lectures and Conversations on Aesthetics, Psychology, and Religious Belief.* C. Barrett, Ed. Berkeley & Los Angeles: University of California Press.

Woolf, V. S. (1976), *The Letters of Virginia Woolf,* Vol. 2. Ed. N. Nicolson. New York: Harcourt Brace Jovanovich.

Yates, S. (1935), Some aspects of time difficulties and their relation to music. *Internat. J. Psycho-Anal.,* 16:341–354.

Yeazell, R. B. (1981), *The Death and Letters of Alice James.* Berkeley: University of California Press.

Zetzel, E. R. (1960), Symposium on depressive illness. I: Introductions. *Internat. J. Psycho-Anal.,* 41:476–480.

Zilboorg, G. (1951), Psychoanalytic concepts of sleep and dream. In: H. A. Abramson, Ed., *Problems of Consciousness.* New York: Corlies, Macy, pp. 62–88.

——— (1967), *Psychoanalysis and Religion.* London: Allen & Unwin.

Zinberg, N. E. (1971), G.I.'s and O.J.'s in Vietnam. *New York Times Magazine,* Dec. 5.

——— & Weil, A. T. (1970), A comparison of marijuana users and non-users. *Nature,* 226:119–123.

Zubek, J. P. (1969), *Sensory Deprivation: Fifteen Years of Research.* New York: Appleton.

NAME INDEX

SUBJECT INDEX